The Overlook
Guide to
Small-Scale
Goatkeeping

The Overlook Guide to

Small-Scale Goatkeeping

by BILLIE LUISI

THE OVERLOOK PRESS
WOODSTOCK, NEW YORK

First published in 1985 by
The Overlook Press
Lewis Hollow Road
Woodstock, New York, 12498

Library of Congress Cataloging in Publication Data

Luisi, Billie.
The overlook guide to small-scale goatkeeping

Reprint. Originally published: A practical guide
to small-scale goatkeeping. Emmaus, Pa.: Rodale Press,
c1979.
Bibliography: p
Includes index.
1. Goats. I. Title. II. Title: Goatkeeping.
SF383.L85 1985   636.3'9   85-8910
ISBN 0-87951-230-X

For my favorite Capricorn, my daughter, Thecla Luisi,
and
in memoriam, Holly Cantine

CONTENTS

Acknowledgments viii
Introduction ix
1   Goat Terms 1
2   Acquiring a Dairy Goat 8
3   How a Milk Animal Works 26
4   Milking 30
5   Breeding 37
6   Birthing and Kid-Rearing 51
7   Feeding 73
8   Shelter and Pasture 83
9   Management and Control 96
10 Well-Being and Its Divergences 105
11 The Goat in the Kitchen 129
12 The Goat in the Garden 160
13 Goat Economics 170
Appendix: Artificial Insemination 189
Notes 191
Bibliography 195
Resources 199
Index 201

I wish to acknowledge the help of Lou Talento, Sue Bean, Chuck Liscum, Ruth Levine, Susun Weed, River Lightwomoon, Justine Swede, and Arya Nielsen in the maintenance of the Woodstock Clay herd and their support through the writing and preparation of this manuscript.

Thanks are also due to Sam Shirah and Benita, Charles Chapin, Walter de Graf, Robert and the late Mary Hersey, for sharing their experiences, old-time advice, and encouragement.

Thanks also to all the Considines at Diamond Farm, Breezy Hill, and Sunshine Farms; and Frank Holder, Bill Marcy, and Gerald Lehmann of Dutchess County, New York.

Thanks for the photographs go to Lynda Suzanne for those appearing on pages 22, 25, 26, 64, 66, 96 and 139-45, to L. V. Talento for those on pages 70-71, and to the Rodale Press photographers for all the others.

Thanks for the illustrations go to Erick Ingraham.

INTRODUCTION

The underlying assumption of this book is that the dairy goat should be both the mascot and figurehead for the soft- or appropriate-technology life-style. She has always been the protein resource of cultures living on marginal land and of lower-class rural populations in times of depression, war, or other periods of transition and unrest. The modern milk goat is an undeniably efficient animal. This has been obscured in Western countries, particularly those possessed of rich land masses and technological resources. Cow dairying became dominant in all such advanced nations and goat dairying in those countries became the province of the cultivated, mon-eyed, and leisured amateur or hobbyist, with occasional up-surges of interest among marginal, economically hard-pressed persons, both rural and suburban, when times grew bad. Goat technology is small-scale technology. Basically, I am fond of imagining avid legions of backyard goatkeepers springing up in suburban enclaves and on the fringes of the great cities, as well as envisioning mini-herds browsing the rural country-side.

Any well-managed dairy animal of good breeding is an eco-logically sound resource, for she converts herbaceous substances that humans are otherwise unable to use into food, fuel or fiber — milk, meat, hair, and skins — and in the process produces manure, nourishment for soil in need of replenishment. What distinguishes the milk goat from the good dairy cow is her capacity for survival and production on poorer pasturages — even scrub if necessary — and the fact that a dairy goat's production is based on labor-intensive input rather than on capital input. The scale of the milk goat's demands and production also define her as a more appropriate animal than the cow for household milk supply and home garden fertilization. In the average household, it is easier to use the one gallon a day a goat yields than the five gallons a day the cow yields. The goat's manure output is similarly scaled to the home garden.

All of the above sounds very abstract and doesn't give the reader a picture of how and why I got mixed up with goats. I was urban born and bred, and only after thirty years left my home in New York City. I had become a potter over those years because I loved clay. I had also found a politically and socially integrated life-style that was congenial to me. As a craftsperson, I worked with my hands with responsive, natural raw materials. I created real wealth by creating handmade goods — often more durable and frequently more beautiful than those manufactured by mass industry. I created my own schedules and commitments and could have channels for self-expression and personal growth at the fringes of the American political and economic system. It was a minority life-style for sure, and it could be a very tight, low-consumption, ethically scrupulous, aesthetically compelling, and politically conscious mode of existence if I chose. In the early mid-sixties many craftspeople lived at the economic margins of American society, but understood the basics of deceptive affluence, exploitation of people and resources, and the falsity of the individualism expressed in new cars and plastic gadgets that constituted the nine-to-five wage earner's share of the pie.

All of this changed in the expansionist late sixties. Many persons who became craftspeople in that decade became very successful in terms of money and recognition, and the affluent life-style and its "rewards" spilled over into the everyday lives and ideological spheres of American craftspeople. Craftspeople began to think that they were part of the professional mainstream and thus entitled to the full rewards of dedicated, hardworking Americans, both in status and in money.

INTRODUCTION

It was a purely subjective conviction of mine that good craft had to maintain some distance from the mainstream of American reward structures and gratifications. I was slowly coming around to a political craft position that was based on de-urbanized, decentralized activity, direct relations between buyers and producers, and creative activity related to broader cycles than the Christmas gift market. After 1968, I began to believe that my craft would evolve further and the quality of my life improve in a broad way if I left New York City. Many people felt the same way. With so many people moving to the country I began to think that the village potter could rise again. It has been the dogma in ceramic circles that the day of the village potter has been over for a very long time. I was coming to the reverse conclusion — that decentralized crafts-people serving local needs and people, dependent on local supply channels and materials, could be the nucleus of still another craft resurgence, one that did not plug into metropolitan stores for outlets and current mass merchandising techniques for value systems. It was a vision of a pattern returning to older patterns, of craft integrated with agriculture, small-scaled, seasonal rhythms, the land, and people's real needs. With the publication of the *Whole Earth Catalog* and its supplements, I felt my visions supported; I also felt such publications would provide the technological backup, communication channels, and information I would need to make the change from the city to country life.

Leaving the city, heart of my heart and root of my being, was my first step toward goat dairying. Of course, at the time, I had no such idea. My decision to begin keeping milk goats came about because I needed manure. One of the *Whole Earth Catalog* dreams I was most taken with was that of methane production, the production of fuel from anaerobic decomposition of wastes. I could be totally self-sufficient — dig my own clay, produce my own food, and derive my own fuel. Keeping goats would provide me with manure for manufacturing methane to fire pots. It seemed so clear and simple, way back there in the city winter of 1969. Naturally, I had no concrete ideas as to how much manure and cellulose-type matter would really be needed to produce sufficient methane for me to fire kilns to stoneware temperatures. By 1971, I was in the country and knew a bit more.

For one thing, I knew that no one in the United States

knew anything or (what was much more critical) had done anything to produce a steady supply of methane to heat a house, run a stove and lights, and fire a kiln to 2,300°F. (1,260°C.). I'd also learned that to produce a sizable amount of methane in a northern climate one would have to have on hand the manure of a large pig or poultry farm or the manure output of an entire cow dairy plus the equipment to haul that tonnage of shit around.[1]

But while doing the research, I had gotten hooked on milk goats. They sucked you into loving them because they were so individual, personable, and damnably clever. And they turned out to be neat, clean, odor-free, and without nasty, unhygienic habits — they were the exact opposite of the bad press they'd received. They did *not* eat any and everything, as my mother had whispered to me. (She put them in the same category as lobsters — exotic and fascinating things that were also dirty, ugly-looking, suspect food scavengers.) Goats did not strip the Mediterranean basin of its foliage or gobble up the cedars of Lebanon. The goats I was meeting ate only the best and most palatable of flora, being superfastidious and tenacious about their canons of edibility. Indeed, their only drawback was the neatness, compactness, and quantitative inadequacy of their shit. A small flock of goats just does not produce enough manure to supply an ongoing methane digester.

Unsuccessful though my research had been, it had introduced me to people as well as to goats. Judging from the people I was meeting in my goat researches, I had somehow hit upon another subculture of disciplined, individualistic, soul-satisfied people.

Speaking from my own experience, I can say that most city dwellers no longer know what fresh milk tastes like, for milk supplied to urban folks today is pasteurized, homogenized, vitamin-fortified, emulsified, at times adulterated. This milk is so different from the real thing that, like bread, it has to have good things put back in. Extended shelflife has also become an important criterion. Milk in its waxed container bears little resemblence to milk straight from the teat as far as taste, texture, butterfat content, and vitamin and mineral content are concerned. I once believed all the educational filmstrips I had watched in elementary school that stated un-

INTRODUCTION

equivocally that modern dairying was bringing me the purest, safest, cleanest, freshest milk in God's creation. Neither these early impressions nor any amount of statistical data produced for me now stand up to the first glass of raw milk I drank.

When I realized how much I like raw goat's milk, I decided to keep dairy goats but forgo methane. I have not yet achieved a village potter's existence, but every year I find myself closer to the little patch of land I landed on. The coming of the goats has made my changed life possible, largely because the goats keep me close to home, curing me of wanderlust and notions of personality expansion based on travel and vacationing. Additionally, the goats supply all the dairy needs for a household of five people. They supply all the fertilizer for our garden; and the garden is now our year-round sustenance — we raise over $1,000 worth of food on the place. The ongoing basis of this abundance is goat manure, easily managed without machines and equipment, and truly fertile. The goats have served to formulate a new focus, the whole cycle of giving and receiving that comes about once one makes a commitment to caring for small stock and the land we live on.

The entire process, from methane dreaming to leading our first milk doe home, from hauling the manure to growing the turnips, has been epiphanic. If, a generation or so ago, E. B. White could find God under the compost pile, I can add my amens and personal revelations, for the divine spark of the mother of us all is there in the yogurt and cheese fermentations, in the steam from the milk pail on a ten below zero morning and in the nuzzlings of the newborn kids searching for their first touch of teat and suck of milk.

One of my recent revelations has been that many new goatkeepers have not found their experience to be a happy, exciting, or illuminating one; that their way was beset with woes, guilt, and disaster. Through a long series of goat happenings, goat gossipings, and calls for help, I began to feel there is a broad lack of information about goatkeeping.

In actuality, there are numerous books, pamphlets, and organizations, all good sources of information. Much of this is bypassed by newcomers because they are bent on learning by doing, have deep prejudices against written authorities, or find much of the literature talking in a strange vocabulary.

Unless there happens to be a knowledgeable old-timer

down your road, you must turn to books after awhile. This book was birthed to fill a gap I have sensed, to chronicle my adventures and everyday goatkeeping life, to pass on some experientially based information, and to draw a picture of the whole that has dawned here since the coming of the goats.

I have been concerned to provide information in an accessible, reasonably priced form that does not obscure or make light of the work, energy, and intelligence a newcomer must bring to the undertaking. So many books err by presenting a roseate picture that novices do not relate to. Crusading articles in the literature of the back-to-the-land movement have a tendency to gloss over the demands and difficulties involved in taking on the responsibilities of caring for small stock. I have tried to fill an access-tool gap by including a chapter of definitions, an appendix, a bibliography, and an index.

Pondside, Woodstock-Saugerties Road
Spring 1978

The Overlook
Guide to
Small-Scale
Goatkeeping

CHAPTER 1

GOAT TERMS

There is a real gap between the veteran goatkeeper, breeder, or farmer, and the enthusiastic beginner. The newcomer's fundamental problem is finding a way through the tangle of unfamiliar language. The veteran assumes that others have a basic comprehension of simple animal husbandry terminology, and sometimes the beginner hardly realizes his or her handicap until after the first goats are in the barn. By then, golden opportunities to learn from the more experienced keeper have slipped away. So that we all know what we are talking about, and are able to ask the right and most pertinent questions at critical moments, I've decided to start this book with a short chapter on goat people's talk.

Goats are preeminently social animals and their social unit is the *herd,* even if the herd is only two or three who do little roaming. Their preferred feeding pattern is *browsing.* Goats don't really graze like cattle who slowly eat their way through whole fields of legumes and grasses. Goats cruise and nibble, eating an enormous variety of plants, brush, weed, shrubbery, and tree tips. They are totally vegetarian or *herbivorous* animals of the *ruminant* group. In other words, they are *cud* chewers, that group of mammals possessing complex, four-chambered stomachs, the first section of which, the *rumen,* serves as a huge digestion vat for all sorts of roughage. There is a steady interaction between the goat's rumen and her mouth. Those tennis ball lumps people see on the sides of a goat's mouth are the food, chewed, swallowed, partially worked on in the rumen, and then sent back up, regurgitated for another round of salivation and mastication. This is cud chewing, a peaceful, thorough, and meditative habit.

The herd is led on its foraging expeditions and escapades by a *herd queen,* usually the eldest or most experienced doe. Mature female goats are *does;* mature males are *bucks.* (Many professional breeders will be insulted if their purebred breed bucks are referred to as "billy goats.") There is a defined social order in the herd, similar to the pecking order found among chickens. Contrary to popular belief, the leading figure in the herd is the herd queen, not the breed buck, imposing though he may be. In fact, the buck is usually not allowed to run with the herd; this prevents unwanted and premature breed-

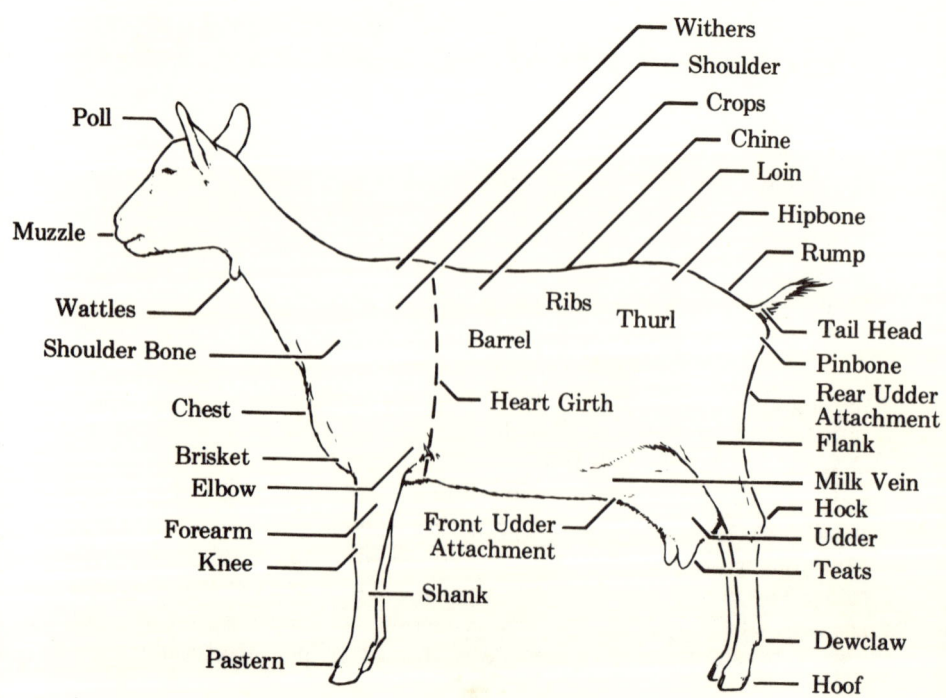

ings and prevents the does picking up the heavy odors of the buck.

Breeding is the premeditated mating of selected does and bucks; it is also an entire science of livestock improvement practice. A *breed buck* is most commonly a *pedigreed, purebred sire* of *proven-out* worth. Proven-out is self-explanatory — such a sire has documented progeny of good dairy character and capability. The term *breeder* usually refers to an experienced goatkeeper who is particularly interested in stock improvement and keeps a herd of carefully selected, registered animals of greater size than a household-oriented herd.

Young goats are known as *kids*. As they grow they advance to be *doelings* or *bucklings*. Castrated bucks are called *wethers* or *wether bucks*.

Does are said to come into *heat* when they arrive at sexual maturity. Technically the heat is known as *estrus,* the period in a doe's reproductive cycle when she is fertile and willing to stand for the buck mounting her. Her heats last, most commonly, about three days, and occur at eighteen-to-twenty-one-day intervals. In northern temperate climates, there is a specific *breeding season,* triggered by the shortening hours of daylight, which lasts from late August to March. The term *rutting season* may be encountered in older books; it refers to the same period. Goats of nonnorthern descent may come into heat any time of the year. However, most breeding talk will refer to "the season." Bucks often know no season, but will stand ready to oblige does at any time — probably the reason the male goat got its reputation for outrageous, constant sexual energy.

Once the doe is bred, she is *in-kid* or pregnant, carrying her young. The *gestation* or development and carrying period is five months or approximately 150 days. At the end of this time she delivers her young, or *freshens*. This is an old term signifying both the birthing of the kids and the does' coming into milk. A milking doe is said to be *fresh* or *lactating,* or starting her *lactation. Letting down* the milk is the term for a dairy animal's starting her milk flowing, usually a conditioned response to feed reward or a patterned time schedule. The lactation is the period during which she produces milk, usually about 305 days in modern goat dairying. The doe's first milk is known as *colostrum*. Colostrum is rich in the antibodies and vitamins needed by the newborn kids to get off to a healthy, vigorous start, and is secreted for up to four days after delivery.

An unpapered, unrecorded, or unregistered goat is a *scrub*. Goats that are of mixed or partly traceable origins may be recorded as a half-grade of a specific breed if one parent is a purebred of that breed. A *registered* animal is one whose birth and lineage has been recorded by a registry organization, most commonly today the American Dairy Goat Association, ADGA.* The *pedigree* is the traceable record of a particular animal. The term *sire* is used in such records to refer to the male parent of the animal. *Dam* is the term for a female goat parent, the mother. The *papers* that people so knowingly discuss in myriad goat conversations are the certificates of registry and recordation sent to the owners of such animals. They are sometimes referred to as the "pedigrees."

Purebreds are those animals whose lineage can be traced back through the national goat registry herd books to the original imported animals of a specific breed designation. The *recorded grade* is an animal with a verifiable written entry in the ADGA registry system. Grades may be recorded as "Natives on Appearance" (similarity to a recognized breed-type) or on the basis of "Performance." Progeny of such grades are not automatically registerable. Does out of breeding-up such recorded grades (or scrubs) to a registered purebred sire can be registered as *half grades* of the breed of the parent registered in the grade registry. Half-grade does that are then bred up to a registered purebred sire of the same breed can have their female progeny registerable as *three-quarter grades* of the breed of the purebred parent. A successive breeding of the three-quarter grade to a purebred sire will result in kids that are *seven-eighths,* and no longer called grades, but designated *Americans* of that breed. ADGA has just changed its regulations on the registration of American bucks; bucks must be fifteen-sixteenths purebred blood of the breed in question to be designated as American Alpine, American Nubian, American Saanen, and so on. Animals registered with an American designation are registered in separate American herd books.

In addition to the papers already mentioned there are several other types of records pertinent to goat dairying. There are star certificates, acknowledgment of advanced registry achievement, and Dairy Herd Improvement Association (DHIA) records. DHIA is a management tool that is administered by USDA through the state land grant college extension service personnel. The keystone of the system is monthly surprise visits by a trained dairy tester who checks enrolled herds by weighing milk and testing for butterfat content. Results are recorded and fed into computers, so that complete documentation of a doe's productivity throughout a lactation is available through monthly computer records. The cost to the keeper depends upon many factors, among them the geographical distribution of the test-

*American Goat Society (AGS) is another registry organization.

er's herds, and the service can be very expensive for the owner of a small herd. However, DHIA records yield so much ongoing, easily verifiable information that even the smallest keepers find it beneficial to participate. This is what goatkeepers are referring to when they say a herd is *on test* — it's being tested and recorded via DHIA. The major means of broadening enrollment in the program has been the development of cooperative group testing, in which small herd owners contribute their labor as a means of reducing costs. A surprise visit is usually paid once yearly by a regular DHIA tester, and monthly testing is done by group members on a rotating basis. For detailed information about putting goats on test, contact the ADGA DHIA committee.

An *advanced registry* (or AR) doe is one that has given 1,500 pounds or more (current standard) of milk in a 305-day lactation. The AR designation is a measure of productivity over the entire lactation. *Star milker* is another designation for an outstanding milking doe. The * on a star milker's papers is earned on the basis of a one-day test and is not a statement of overall productivity throughout the lactation. Information and further detail on these terms is also available from registry associations. Bucks can have stars (*) on their papers, derived from the stars earned by their dams or other maternal progenitors. A new category of records that is being developed is based on classification.

Classification is a term that will be encountered frequently in the future of goatkeeping. It refers to a scorecard-type system that places goats in classes such as excellent or very good, based upon general appearance, dairy character, mammary system, and body capacity. Each of these categories is allotted points, such as mammary system, 30, body capacity, 20, with the total number of points giving the goat her general classification score. The system is similar to that developed by Holstein breeders over the years. Now through computerization and recordation by the ADGA, the classification number will become part of the record of registered animals. The ADGA is encouraging owners to have their animals classified, and currently has three ADGA-employed classifiers in the country. It will eventually be possible to make breeding choices by matching up the scores of classified animals and trying to improve each of the major category areas. Classification relates dairy goats to an agreed-upon general standard of excellence. It differs from show judging in that show judging compares just those animals present at a particular time and place and within a specific group to one another, and then selects the most distinguished from the group at hand. If the group at hand is composed of does classified in the 88 to 90 range, the most distinguished of them would be an excellent animal indeed! If it is a group that has not been classified or has low classification scores, the best of that group is not necessarily an excellent animal, but simply the best of that group. Classification scores are bound to become an important tool in selection, breeding, and goatkeeping generally.

All these designations and papers sound confusing at first. It's easiest to deal with it all by keeping the purpose of papers and productivity records foremost. The fundamental purpose of records is to enable people to identify, trace, or document productive livestock. Take what you need from this apparatus when you are ready.

Modern milk goats in North America usually belong to one of five well-defined breed-types:

Saanen evolved from centuries of selected Swiss breeding stock; they are a large-framed, most commonly all-white, heavy production breed.

French Alpine derived from the importation of a selection of French Alpine dairy animals in 1922. They appear in many colors and combinations, and are also a heavy milking breed.

Toggenburg derive from Swiss and German stock, are often smaller animals than Saanens or French Alpines, and are distinctively marked with white facial blazes and leg and tail markings against a brown coat. They are very copious milk producers. The all-time record holder for production in American goat dairying is a Toggenburg.

Nubians descend from the crossing of British and Eastern-derived stock. They come in all colors and have large pendulous ears and Roman noses, and are famous for high cream content in their milk.

La Mancha is a new breed being currently developed in North America, characterized by smooth, short coats and "gopher" ears.

Most goats of both sexes are born with horns. There are some naturally hornless goats, but most of the ones that seem hornless have been dehorned at a very early age. Most dairy goats are artificially dehorned or, technically, *disbudded,* the horn buds being removed sometime within the first week after birth. A goat without horns is called *polled.*

Goats are exceptionally healthy and hardy animals, but there are a few health terms you will encounter often. *Brucellosis* and *TB* (tuberculosis) are diseases of dairy animals that can be milk-transmitted to humans. Goats should be routinely tested for both once a year. *Mastitis* is an inflamed condition of the mammary glands — the goat's *udder* or bag. *Edema,* a swollen, fluid-retaining condition in the tissues, is often associated with mastitis conditions. For further specific terms and problems of a health nature, consult a good veterinary guide and your local veterinarian.*

---

*My standard health references are C. E. Spaulding, *A Veterinary Guide for Animal Owners* and H. J. Heidrich, and W. Renk, *Diseases of the Mammary Glands of Domestic Animals.*

CHAPTER 2

ACQUIRING
A
DAIRY GOAT

The coming of Woodstock Clay's Mackenzie, Eleanor Roosevelt Toggenburg, and Bela Abzug, doeling, resulted in my personal metamorphosis from a *Whole Earth Catalog* fantasizer straight from the big city, into a down-home, day-to-day goatkeeper. It's a prolonged tale that I hope inspires — and instructs — you.

The last place around this town you would expect to see farm livestock for sale is at the annual Library Fair. The fair is our one big charity extravaganza for the year. It is a crowded gathering of friends, an all-day entertainment, a fund-raising event, a book sale, a plant sale, a rummage sale, a scene of puppet shows, steel bands, mummers, strummers, and mimes. Now and then there are some animal visitors, usually small ones brought to the midway for children to play with and pet; a thin reminder that this rural-suburban township once had its share of farmers and foresters.

I went to the Library Fair the last Friday of July 1972, and found a young pair of Saanen-type kids tethered out on the grass. They weren't there for visitors to pet; they were for sale. Many curious questioners and watchers appeared throughout the long day, but to most folks the pair of fluffy white kids seemed just another fair attraction. Late in the day, I made the owners an offer of fifteen dollars for the doe kid. It was the only real offer of the day and they accepted. We arranged for a visit that night so that I could see the doe's mother. I rushed home to spring the news to my land partner, Lou, and to put together goat accommodations in our projected goat shed. We engaged in convoluted and circular conversations about where and how to find another goat for herd company. It seemed best to pick the kid up several days later, so that we could hunt up a companion goat and finish the stall area.

The goats were living at a very mountainous, secluded homestead site in a last bit of the Bearsville wilderness. Their pasture was woodsy, stony and verdant with a good crop of poison ivy. The goats were sheltered in a lean-to, and all looked hardy and healthy. They had real dairy character and conformation, too. The doe kid's dam was a big, determined-looking Saanen-type doe with a huge, uneven, pendulous bag. She had a capacious barrel, frame, and udder, and had enormously developed milk veining. Her udder muscles were totally weakened and the bag hung lower than any goat udder I had ever seen, but it was soft-skinned and had no lumps. Her stall mate was a big, but slender, chocolate-colored, Toggenburg-looking doe with the classic white facial blazes and other light Togg-type markings. She looked like an elegant antelope, and had a skittish, maidenly temper. Then there were the two kids out of the big white doe. The young twins were a buck and a doe, about seven weeks old. We arranged to pick up the doe kid the following week.

Lou and I talked goaty things all the way home — we were filled with projects, advice, and goat gossip. We knew that bringing one lone kid home was going to mean trouble, so we covered many miles in the next few days looking for herd company — even a donkey. Goats are herd animals and need company. If they aren't provided with goat company, they demand people company. But finding a goat out of the peak spring birthing and selling season turned out to be frustrating and finally futile.

The herd-companion problem had not been worked out when pick-up day rolled around. We wrung our hands and scowled, but in the end Lou went for the doe. We were figuring that we could survive a week or so of morose goat calls and stall-sitting. Then two hours later we found ourselves possessed of an entire herd.

When Lou arrived at the goats' owner's homesite, he found Sam (the owner) in a state of goat overdose. Sam's land partner had disappeared over the weekend. Sam was upset and tied down with the four goats plus a dozen chickens and ducks, dogs, and a garden. He wanted to know if we would be amenable to boarding *all* the goats for a few weeks. From worrying about the loneliness of a solitary kid we were plunged into major anxiety about how to take on the whole group. I felt it was the destined, convenient resolution to our problem, and the experience of having an actual milking goat to relate to really appealed to me. We decided to ad-lib the stall construction as best we could, covered the floor of our van with a thin layer of fresh hay we picked up along the roadsides where the state had just mowed, and rolled off to meet our animal husbandry destiny.

Sam gave us his remaining feed, the goats' tethers, as much advice as he could muster under the circumstances, and good wishes. The two older goats hopped willingly into the microbus. The kids jumped in after them. I held the kids in my lap in the front seat of the bus, vainly trying to keep them from their dam. Sam had said they were recently weaned but still trying to sneak a suck or two when the opportunity arose. Our menagerie gaily covered the seven or eight miles to home with no catastrophic incidents — no wails, evidences of bad temper, or terror of changes.

It was hard getting the older goats to come out of the bus, but we finally lured them out by carrying the kids off to the barn. Our thrown-together stalls (planned originally for one or two young kids) couldn't withstand the curiosity and assertions of the big goats so we tethered them in separate areas of the barn. The kids were very resourceful in finding ways to the teat. We realized that, even for a few weeks' stay, we had to create stalls and feeders that were sturdy enough to house everybody separately, particularly to separate the kids from their dam, and to provide individual feeding spaces for the big goats. We got a first-class demonstration of why experienced professional goat breeders keep only dehorned dairy goats.

All these goats had horns and were really practiced at using them on each other. From what we saw, their butting tactics were mainly defensive measures to protect food or territory for themselves. But the milking mother, Scapegoat, butted people with her horns — particularly little people. Goats, reputedly, did not like changes, and she was busy letting us know that she didn't like having her life and territories reorganized. She didn't much like the separate stall arrangement either. We were insistent on that issue, and in a few days both she and Eleanor settled down in their own stalls and worked out their fears and food competitions by butting the divider wall between them. We had to reinforce the latching several times and rebrace Scapegoat's door.

ACQUIRING A DAIRY GOAT

Given her size, our inexperience, the horns, and the eager, half-weaned kids off in the background, milking old Scape turned out to be a major production repeated twice daily — not quite the idyll projected by the pamphlets and homesteading literature. My only prior milking experience had been two months before at the barn of two professional goat breeders. I had concentrated on getting the right kind of finger circle and closing/opening motion, and on not pulling down on the teats and udder. I had managed to milk a few squirts after several tries. How patient even-tempered, and milk-stand-trained those good goats were! I had never imagined a goat that wouldn't patiently eat her grain while being milked. No amount of reading or barn-visiting prepared me for my first eyeball-to-eyeball confrontation with the great horned goat I now had in the barn. Nothing in her prior experience had prepared her for a locking milk-stand and a solitary, stubborn milker.

Gradually we became more evenly matched. My first gain stemmed from tethering the kids out of sight and out of earshot all day. With their all-seeing, all-hearing, formidable goat matriarch out of the situation the kids eagerly awaited my coming with their pans of milk. Pan feeding them all day made for a lot less background noise and fewer furtive attempts to find teat back at the barn when evening came around. With the kids less hungry, old Scape had to let down for me when I showed up with the pail. Goats are famous for the amount of control they have over their letdown. She grew very anxious to have her grain ration on time, too, even though it meant hopping up on the confounding milkstand and getting locked in.

The last step was to train her away from loosing her final bag of tricks. It wasn't just one bag. She had bags and bags of tricks: abruptly raising one leg and plunging it into the bucket; shaking, all of a sudden, loose hair, hay, or a dead fly into the bucket; letting fly a load of goat berries when the pail was almost full (one or two almost always scored). All of them had the consistent result of spoiling the milk. Last but never least, was the front-ender. The front-ender was infinitely varied: some nights, when the pail was nearly full she would suddenly rear up and, in a reverse rodeo trick, stand on her front legs, letting fly the rear ones helter-skelter. This trick was most successful when the pail was knocked over, the milker shocked into total confusion and the milk completely lost.

Scape's front-enders were a special nighttime routine. She sensed my tiredness and the seconds of wandering attention. And then she would do it. Some nights I got kicked; others, I got milk. For a while I chanted to myself to make sure that I paid attention. Often, I sang while I milked. Some nights these remedies worked, and some nights they failed. The morning milkings were generally easier. The goats knew they were going out to pasture and were willing to play our games in order to get out.

One afternoon I was browsing in a bookstore in town. There I found in my hands the only book of its kind, the goatkeeper's bible, David Mackenzie's *Goat Husbandry*. "Oh yes," the bookseller said, "you can buy it. It has an order slip, but that was from a half-year back. I got stuck with it when the buyer found out it was a seventeen-dollar goat book." I wanted it. I had read it before, but at that point I hadn't been willing to spend that kind of money for a goat book, even though serious goatkeepers had recommended it as the one book to have on the shelf. Yet, there was Mackenzie speaking to me on the first page:

> *Nearly twenty years ago, I retired into a converted hen house with a milking pail, a book of instructions, and an elderly goat of strong character. There was milk, among other things, in the pail when the goat and I emerged at last, with mutual respect planted in both our hearts. The book of instructions was an irrecoverable casualty.*
>
> *No book is a substitute for practical experience; but books become more important as animals become more productive . . . each class of farm livestock requires to be accompanied by a comprehensive work of reference as a passport to the modern farmyard.*

The balanced prosody of a retired Scottish schoolmaster taught me how to deal with my outrageous milk goat and put me on the right track to goat dairying. I had been approaching the whole experience with the nebulous proclivities of the contemporary natural-life-style cultist. But he had met me and my ilk years before. The book let me know I wasn't alone in feeling guilty about taking mother's milk out of the mouth of her babies.

That night, purged of guilt and filled with new determination, a few insights into the goat psyche and some tricks of my own, I advanced upon the goat shed. I told old Scape that she was being renamed — from that night on her name would be Mackenzie. In the course of the milking she tried a front-ender. This time I grabbed an ear and her tail and pulled them toward each other, hard. She resisted and got one rear foot into the milk bucket. There went half the day's production. I picked up the pail and tossed the remaining fouled milk over her head. The upside-down pail sat over her horns, and milk dripped slowly down her face. It was an outrageous baptism and the final front-ender. She thenceforth not only stood to be milked, but she calmly produced ten to twelve pounds of milk a day for the next three months and then went on to a respectable extended lactation. I bought her from Sam and for twenty consecutive months was the recipient of the best tasting milk I had ever had.

Advice for the Beginner

It has now come to pass that I average ten to a dozen goat calls every week, many more during breeding and kidding seasons. For seriously interested persons, my advice runs as follows:

— Read widely in books and pamphlets.
— Visit and see as many goats as you can find to look at *before* buying anything.
— Buy your first goats from the most experienced goat breeder nearest you, if the breeder has any female stock to sell.

The books may cost you more initially than a freebie goat or a goat from an auction but they will save you trauma and heartache. A reliable breeder has a reputation to maintain, an interest in your using pedigreed bucks for servicing the does you buy, and other commonsense reasons for being straightforward. You can trust professional breeders. You cannot, however, expect such persons to be selling the best does of their herds. Breeders have a commitment to improving stock lines generally by breeding what's best and strongest. You're not going to be able to buy the best proven-out does — they are the ones the breeder keeps. What is usually available is either culls or young solid stock that will not be the end-all of the show ring or set productivity records. Culls from a great line may turn out to be far stronger and more productive than goats of unknown origins.

Be prepared to make substantial financial outlays if you choose to buy registered stock from a well-established breeder. Visit around as much as possible; see goats till you are counting them in your sleep. It will help you determine what, where, and how much to spend, and to settle those nagging questions — registered or not, what breed, how many. One of the major decisions will be whether to spend the extra for registered animals or whether to buy scrub.

Scrubs are among my favorite goats. Before professionals take offense, I'd like to explain this preference. Scrubs form a subculture in the milk goat world. Most of the dairy goats you'll encounter in visiting the various herds of your area are registered goats, descended from recorded ancestry. Scrubs are any domesticated but unpapered goats — goats of unknown origins, social nonentities. I am fond of them because they have taught me much about goatkeeping. Most scrubs will live up to the negative implications of the term: small-dimensioned, ratty-coated, unproductive animals lacking true dairy character. But there are other kinds of scrubs, and I find the term no longer totally pejorative. It is partially a bias derived from my personal experience — all my first goats were scrubs — but, I now have realized that many registered goats are not necessarily superior to some of the nonentities. Often, a solid line out of good original stock has been maintained by a knowledgeable goatkeeper who has no interest in showing, registry, or other forms of recording, just an interest in milk production for the immediate household. Unpapered goats out of such situations are technically scrubs but can turn out to be very productive animals. And unfortunately, registry alone does not assure productivity, not under the current American system of registering every goat a breeder chooses to record.[2] Take a good look at the scrubs as well as at the registered herds in your vicinity. Learn about dairy character — not just how to read papers. Formulate some guidelines so that you can have a firm feeling when decision-making time comes knocking.

Planning the Acquisition

There are numerous schools of considered thought on the general subject of acquiring a dairy goat. Most people concur that the goatkeeper-to-be must realize that two goats are a necessity. If a person insists on keeping only a single goat, there are great opportunities for mutual unhappiness in the offing. Goats need a social environment, preferably a herd environment, and have no compunctions about making you the herd. Bring home a single goat and you will become a herd substitute. It is the solitary goat that hassles; she hassles you and the neighbors, marauds in the plantings you want her furthest from, bleats and blathers for company, and eventually turns you into a human herd companion.

Most people don't envision a pet venture when they start fantasizing about a goat nibbling out on the lawn. They are thinking about a household milk supply. If, as many folks find, two goats are not available, it is possible to start off by keeping a single doe and some other herding animal for initial company; a young castrated buck, a donkey, horse, or pony. For someone on a tight budget two does are far and away the most preferable start. There will be the advantages of increased milk and staggered lactation periods for about the same amount of labor and human energy as would be expended on one goat. Three or four lactating does define a new enterprise. The feed bill increases, but it is human energy that is the greatest expenditure in goatkeeping. I found that the energy input jumped enormously when we arrived at the point of having three productive milkers in the barn. Chore labors (feeding, watering, haying, mucking out, milking, hauling hay and grain) took a great leap then.

People who call me about getting started with goats always ask whether it is better to start with goats already milking or to buy doe kids and raise them to productivity. Theoretically, it is a sweet problem and advantages can be found from both angles, but I feel this is largely a rhetorical exercise. The current realities of goat raising do not allow beginners many choices at all. You start with the stock you can *afford, locate* and *house.* Good producing milk does are hard to come by. Don't decide beforehand that you will start only with kids or only with goats already in milk. Stay loose enough to buy whatever *good* goats come your way.

I'd been interested in goats for about four years before I bought any. When we bought our acreage, we purposely chose a place that had sound existing outbuildings. I knew that raising one's own goats and chickens economically often turns on the availability of shelter. If one must face the prospect of spending a great deal of money to build housing for homestead animals, one either puts off getting animals, or goes into debt for an activity that does not realize profits in the ordinary marketplace sense of the word. The plain facts are that raising *any* animal to productivity costs money not only for initial stock and breeding, but for feed (whether you buy it, raise it, or barter for it) and requires time and human energy for the careful raising of the young.

By July of 1972, I had seen more than 100 goats, but only one that I had wanted to buy. She was a good-producing, purebred Saanen who was being sold because she threw kids with a discernible tan stripe down the spine. If you show purebreds you could not show a Saanen with a spinal stripe or take orders for kids out of that line in good conscience. I did not buy her because I had made a decision not to spend heavily for my first goats. I not only had no prior practical experience, but I didn't want to sacrifice a fine animal to my own inexperience. Once I became aware of current prices for good registered stock in my region (the Northeast), my goat fantasies were appreciably revamped. My region is an area of high overhead for feed and hay, and goats could easily become a financial drain as well as a drag in terms of labor and commitment, particularly if the first ones were initially expensive and then, through my inexperience, things didn't work out well.

By the summer of 1972, I knew that there were few inexpensive milking goats for sale. Six-week-old doe kids with purebred or American designated papers started at seventy-five dollars and went up and

(continued on page 18)

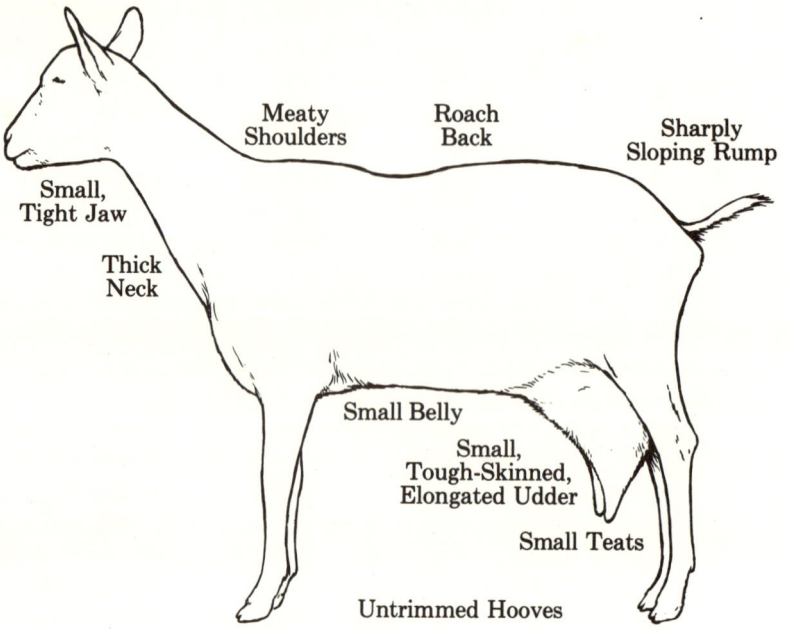

Meaty
Shoulders

Roach
Back

Sharply
Sloping Rump

Small,
Tight Jaw

Thick
Neck

Small Belly

Small,
Tough-Skinned,
Elongated Udder

Small Teats

Untrimmed Hooves

A Goat of Poor Dairy Character

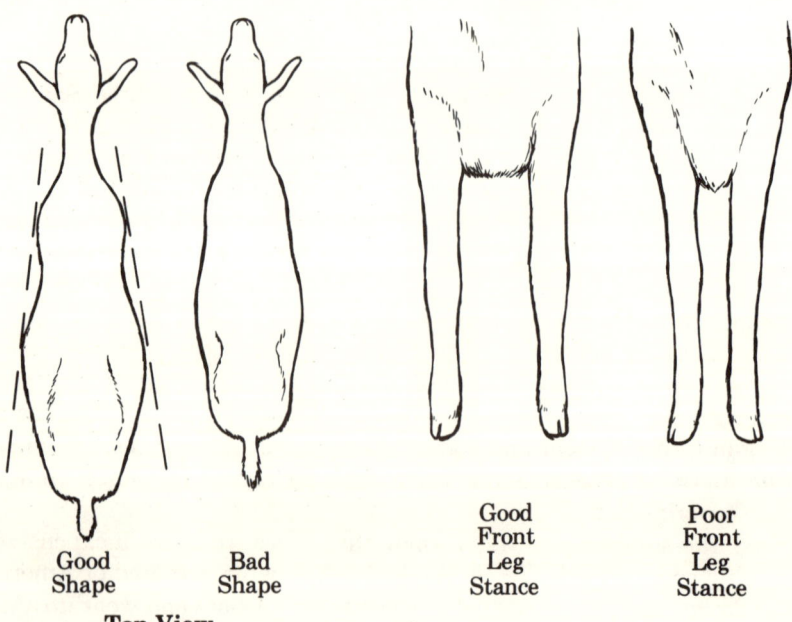

Good
Shape

Bad
Shape

Good
Front
Leg
Stance

Poor
Front
Leg
Stance

Top View

16

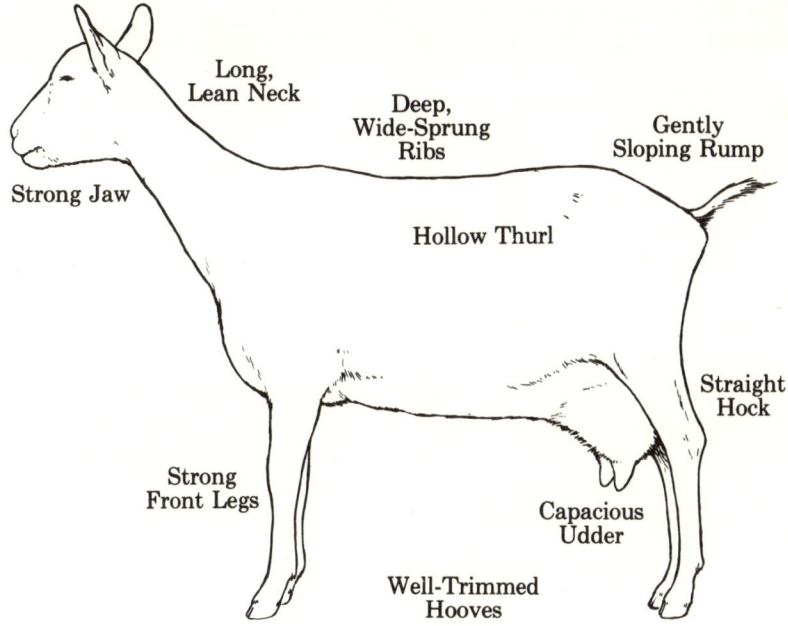

A Goat of Good Dairy Character

Seeking out dairy character—
1. *Alert and angular, with a long, lean browser's neck.*
2. *Good-sized head with a strong jaw, large and open nostrils.*
3. *Long and fairly level rump, but not absolutely straight.*
4. *Large, roomy pelvis.*
5. *Even, well-hung, and capacious udder.*
6. *Well-formed, sturdy legs; wide apart rear legs that allow ample udder room.*
7. *Feet, square and compact with a level-looking sole; pasterns should meld into hoof smoothly.*
8. *Long, wide apart and well-sprung ribs; large barrel capacity.*
In cold climates, look for a large barrel that indicates maximum capacity for heat generation through digestion of great amounts of roughage. In hot climates you may want a goat with a less wide, more slabby conformation. When planning to manage goats with a stall system and exercise lot, there is less demand on legs, feet, and udder musculature. Goats that range and browse for most of the year have to be selected with attention to extra sturdy legs, feet, and udder attachments, and should have very developed jaws. Study the parts of the goat diagram and try to visit both ranging herds and herds kept in more confined circumstances.

up. Recorded grade doe kids were fifty dollars or more. Folks who had good-producing, unregistered milkers were not selling them. How did one acquire a good first animal or two? I found no acceptable way. Although friends bought goats at auction or from herds being broken up or severely culled, and some received gift goats, it seemed, on the whole, that good goats were not auctioned or given away. Poor ones were. I just kept looking.

The market is now tighter than a few years back. All the costs necessary to bring the young doe into productivity have zoomed. The inflationary impact of sky-rocketing fuel and grain prices has converged with an upsurge of interest in dairy goats, and we now have the most extraordinary and soaring stock prices in American dairy goat history. Some persons speculate that real hard times may drop milk goat prices. That remains to be seen. Kids from outstanding herds are very costly today. If you decide to buy registered stock, be prepared for this new situation. Read the figures in *Dairy Goat Journal* over the last two years — they may shock you if all you've read about goat economics is articles in *The Mother Earth News* or other homesteady publications where occasional authors still tout $50 composters on the hoof and miracle milkers for $100. Remember that these publications are crusading to get people back to the land with small livestock. They tend to be positive, enthusiastic, and vague or out of date regarding hard-core money matters. But on the balance side of the financial picture, consider that a good registered goat costs much less than a registered large dog, returns a solid protein value for your feed dollar and care and provides a useful manure for your garden. She is a classic example of the ecological value of keeping small productive animals instead of pets.

Watchful Waiting

If you have no prior experience at keeping livestock and/or little money, you can still start goatkeeping. The basic tool for persons in this situation is watchful waiting. Work toward goatkeeping by:

- putting together a suitable shelter.
- visiting goatkeepers, seeing as many does as possible and talking with as many experienced keepers as is feasible — they are your resource people.
-learning to milk and how to tend to basic chores such as watering, haying, feeding, hoof-trimming.
- going to goat shows and goat club activities.
- studying selection criteria so that you know good dairy character when you meet up with it.

— paying attention to show records, *Dairy Goat Journal,*
the ADGA handbooks; even if you do not plan to
show, this familiarizes you with the major herds, breed
types, show and production standards and recording
methodology — all indices of good milk production.

The prospective goatkeeper who does the most homework prior to
purchasing dairy goats will be the one to make the best selection
decisions. Just getting "papers" with your goats is *not* a guarantee.
However, buying registered animals is theoretically not as chancy as
buying unknowns. I use the word "theoretically," with some emphasis. Getting started with registered stock is an involved undertaking,
costing more but granting no assurances of more milk in the pail for
the costs. If you choose to start with registered animals, write to the
American Dairy Goat Association or the American Goat Society, and
join. From these organizations you will obtain the criteria and procedures for registry — this is information explaining what the "papers"
mean. You will receive the annual handbook recording the doings of
all the top dairy goats — the breed leaders — their test records and
show wins, and the directories to the established professional breeders
in your region (and the nation). It is the buyer's responsibility to be
familiar enough with this material to interpret papers accurately and
select stock in an informed manner. When purchasing registered
stock, ask to see production records on the dams, granddams, and
sisters of your projected purchases, as well as the pertinent registry
papers. All these papers are your guide to the potential of an animal,
and are an indication of how much milk there has been in the family.
If no records other than registrations are available, you shouldn't be
paying top prices. It is *your* responsibility, however, to learn how to
read those papers you are paying extra for. If you have had prior
experience with other productive livestock, have some acquaintance
with the paper convolutions, can supply a reasonable amount of initial capital and time, and have accurate hay, feed, and pasturage
projection figures, it is worth your while to start with the best registered does you can locate. In the long run you will come out far ahead
of folks starting with fair scrubs, gifties, or auction animals.

Start Cautiously, Build Slowly

For those of us recently come out of the urban canyons who are
attempting to homestead with low overhead, who have had little or
no experience with real barn animals and who want a household milk
supply, it is probably best to start as inexpensively as possible. That
translates into buying scrub goats. There is an underground of house-

hold-oriented goatkeepers, folks who keep a couple of goats for their own table supply. These people don't show goats, don't register goats and don't keep records of the milk yield or the progeny. They chuckle a lot at current tales of thousand-dollar milk goats. They breed their does to the nearest bucks regardless of breed or milk background considerations. Some of these keepers have happy, healthy, productive animals. The best way to tune into this network (which is not very visible until you get out and look) is to put out the word about your interest in milk goats to gardeners, farmers, people who raise chickens, plant growers, and feed or hay suppliers. Then comb your local newspapers and especially pennysaver-type papers regularly. Keep your eye peeled for goats out on the browse during grazing seasons. You can usually spot these animals if you are passionately interested and observant. Very few small-scale household keepers maintain their goats as stall animals because it is too expensive a method of management, requiring costly grain and hay feeding year-round.

If this is how you plan to start dairying, you need to know even more before you buy. The person buying registered animals can rely on the breeder, and pay a tariff to compensate the breeder's experience, knowledgeability, and scruples. When buying goats of unknown origins, it is all up to you whether you come out burned or with milk in the pail. Do double the homework and prior looking. Try to locate solid, producing goats and keep an eye out for all their doings: their production, breeding, and kidding patterns. If you buy a mature doe or a kid from a household herd, use all your common sense. Check out the condition of the dam, taste her milk, look over the shelter and pasture, ask about the grain the animals get. Find out what kind of sire was in the background. If he is local, see him and his local progeny. Compare the goats and kids up for sale with the ones not being offered. Look at the animals' feet to see if the hooves are trim. Judge the condition of the animals' coats. If the doe is in milk, look over her udder closely. Feel it for lumps or hardnesses. Check about for scabs or other skin irregularities. Milk a little from each teat into a small cup and look at the milk for signs of clotty, flaky, or streaked milk. Many older goats have poor udder musculature, and the bags on some four- or five-year-olds (check their teeth for age verification if you can safely open their mouths) are sagging and uneven. These are does who have always had kids on them or have had rough early years. Unless there's visible damage, hardnesses, an extra teat, or signs of mastitis in the milk, don't be totally put off. A loosely hung bag may still be the udder of a decently yielding goat. My Mackenzie has a terribly hung udder that would make a show judge faint, but still she

20   *(continued on page 25)*

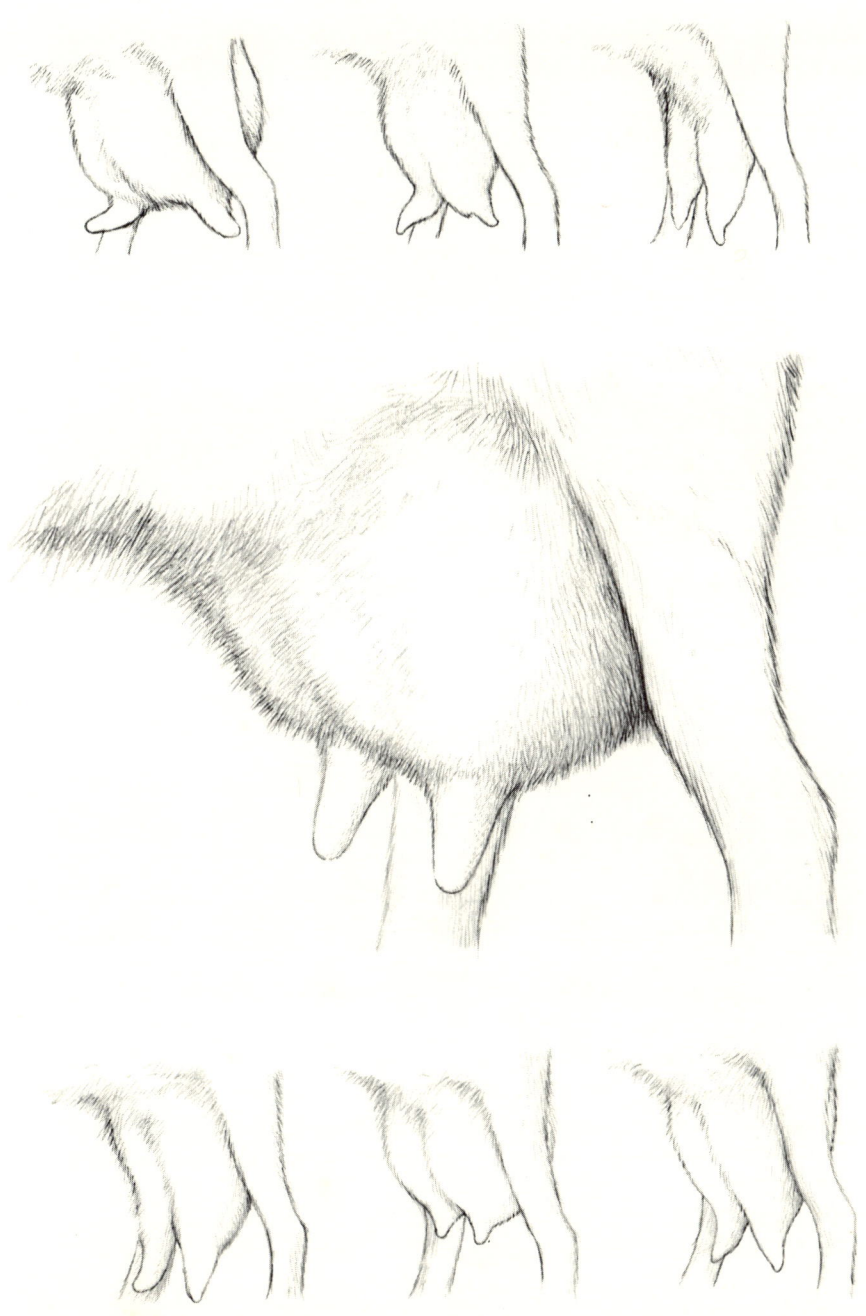

An example of a well-formed udder
and six examples of shapes of ill-formed udders.

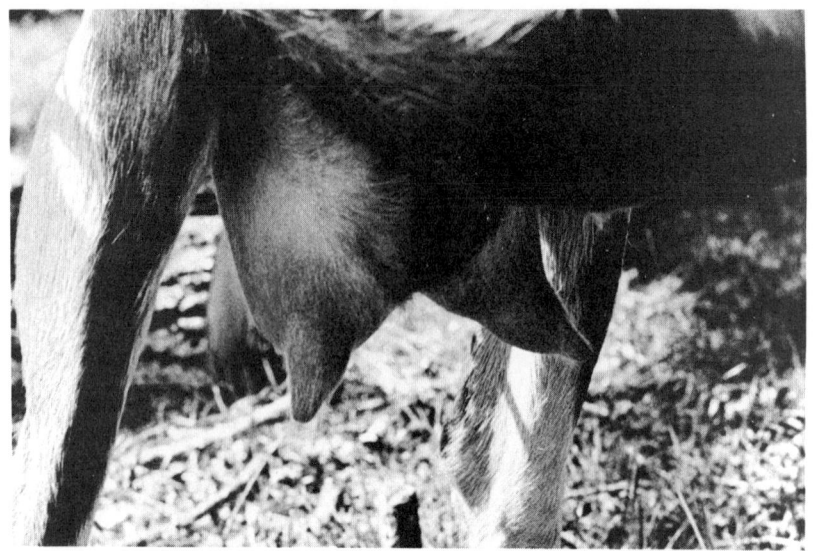

Udder of a goat turned out to browse after the morning milking. It is a high, tight, well-hung first freshener's bag at work—not glamorized or "bagged up" for the camera.

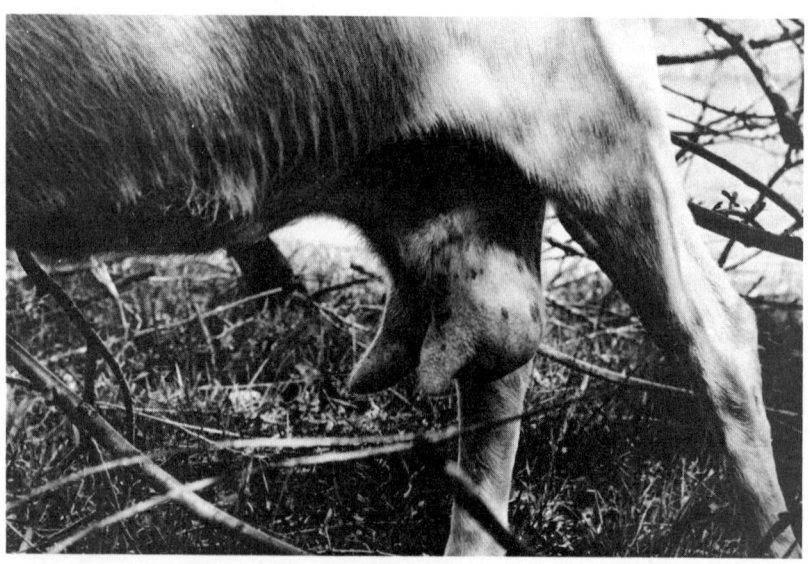

Another early morning, post-milking udder. Notice the complete lack of musculature, probably a result of her owner's allowing triplets to nurse for twelve weeks.

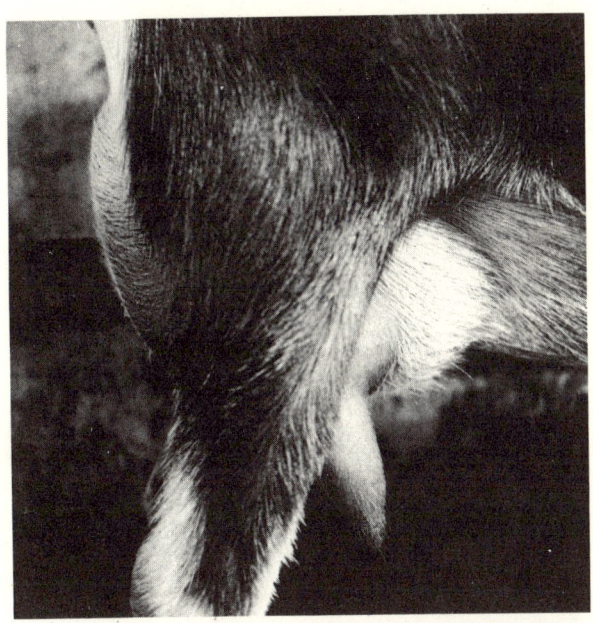

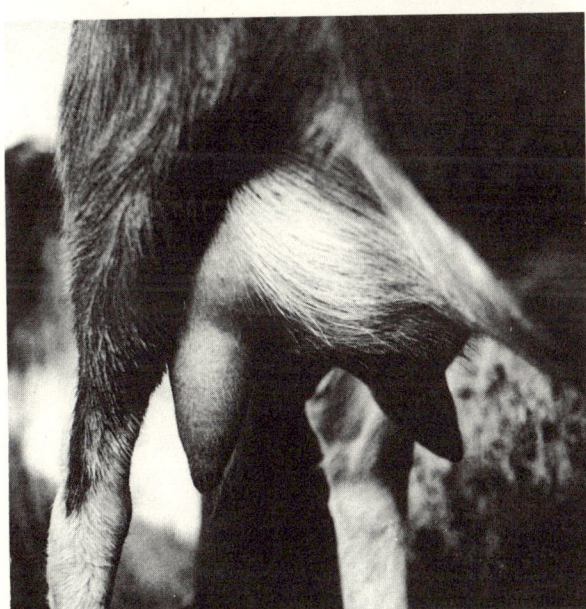

Check all views of the udder before and after milking. This particular bag should have some of the excess hair trimmed from it for sanitary reasons.

Check rear musculature to see if it is strong and taut. The best way to do this is with your hand.

This is a good, capacious, well-hung bag. The teats are well-defined and easy to milk.

A doe with her weaned kid. She has an even, high udder between decent legs. Check out does you are thinking of buying when they are out on pasture so you get a good look at bags and legs when the does are at work browsing.

yields eight to ten pounds (four to five quarts) at the seasonal lactation peak. Make a reasonable offer in terms of your locality and personal finances for any suitable doe kids or available milkers you locate, and take them home for starters.

You can improve the milk yield of inexpensive scrub starter stock within three years by taking care to breed your first does to solid, registered bucks of the particular breed you are most interested in. Make sure the bucks have milk in their lines; this means interpreting more papers and getting about to visit the buck's offspring to see what they are producing. You can't go by a buck's looks alone or some distant ancestor's show wins. Check chapter five for more details on this method of increasing your herd's future milk yield. It is a much slower route, buying scrubs and breeding up a herd from nonentities. However, it may be the only method of getting started in these times of high feed costs, fast disappearing pastures, and heavy demand for dairy goats. It is best to get started.

25

CHAPTER 3

HOW A MILK ANIMAL WORKS

A friend asked me to come to a local community college and give a talk on goatkeeping. It was my introduction to the disconnectedness of people's thinking concerning the foods they eat. How fundamental foodstuffs came into being was a matter of mystery to the class. No one in the group knew what made a goat a milk goat. They didn't know it had to be born, be raised up from kidhood, become mature sexually, be mated, conceive young, give birth healthily and feed her kids, before the goatkeeper (and the human consumers) could have milk in the pail and a dairy animal in the barn. Words such as "rumen" and "lactation" met blank gazes.

Thanks to my high school biology teacher, Mrs. Fritz, these plain facts of mammalian life had been clearly and irrefutably impressed on my consciousness. The biological truth had been hammered home that these same facts applied whether the mammal in question was a human being, a cow, sheep, horse, or goat. If you are seeking milk, you either must resort to the reproductive cycle of mammals or start soaking soybeans. I had always assumed that other city folk, and certainly all country folk through their direct experience, were aware of this fact. Either the passings of the Mrs. Fritzes of the urban schools, curriculum changes, or the triumph of supermarket food ideology — or all of these combined — have created a huge gap in food consciousness in the minds of many Americans. It is unhappily common to find would-be goatkeepers of both rural and urban backgrounds muddled about how a milk animal works.

Ruminants

The dairy goat belongs to the ruminant family. Commonly, a ruminant is defined as a cud-chewing mammal, which is accurate but not to the point. The milk goat's physiology includes a four-chambered digestive organ or stomach. The chambers are called the rumen (or paunch), the reticulum, the omasum, and the abomasum (I like to stress the *rumen* in *ruminant*). When your dairy goat is chewing her cud, she is rechewing roughage and all sorts of crude fodders that she has already eaten and sent on to the rumen. She regurgitates this food and is happily reworking it when you see her moving those tennis-ball-like lumps around her mouth and jaws. Taking the emphasis off the cudding and putting it on the rumen itself is the key to good goatkeeping. The goatkeeper who ignores the goat's rumen is like the gardener who ignores the condition of the soil.

The goat is not a cow or sheep. Among ruminants of her size, the dairy goat possesses an extraordinary rumen. She can eat twice as much as a sheep of comparable size, and she'll eat tough, crude and fibrous plants that a dairy cow won't look at. She can survive and produce on the most unusual grazings and feeding.

The supply of fibrous materials gets sent into the rumen; the digestive working over of it provides warmth and the fundamental materials needed for growth, maintenance, and eventual milk production. Microorganisms abound in the paunch or rumen and generate heat as extremely tough, coarse, foodstuffs are broken down for further digestion in the goat's system. The bacterial fermentation produces warmth as well as food. Consequently the goat can withstand rough, cold, wet, or changeable conditions as long as her rumen is well supplied and active. A properly developed rumen enables the dairy goat to perform her personal miracle — the conversion of vast amounts of scruffy growth and weed, poor hays, and relatively small amounts of grain into a quality protein food.

The proper development and capacity of the goat's rumen also relate directly to her mature milk production potential. It is important to remember this fact while rearing doe kids. If she keeps generating warmth via her built-in heating plant, her rumen, the milk goat will not succumb to chills or pneumonia.

Lactation

Like all other female mammals, a doe must undergo the changes leading to maturity and conception and must carry her young to term (in goats approximately 150 days), and healthily deliver her young before she comes into milk. The milk is produced as food for the coming kids. But we are not robbing the young of their vital sustenance when we milk, because modern livestock has been improved to yield greater amounts of food than the bare amounts originally needed for nursing the young. We also extend the milking period beyond the normal nursing period found among feral does.

The time period during which a doe produces milk is called her lactation. All the "lac" root terms repeatedly encountered in dairying literature stem from the Latin verb *lactare,* which refers both to nursing, or giving suck, and to containing milk. A 305-day, or ten-month lactation is the common goal in contemporary goat dairying, but the length of the lactation is controlled by the needs and orientation of the dairy farmer. Solid, high-producing does can be run for extended lactations for purposes of resting the does from pregnancy stress, of alternating milk supply, or of staggering freshening schedules. The ten-month lactation is not a physiological rule, but it is the most common practice in North America because the income from selling kids yearly is very important to most keepers. Goats in warmer climates, such as the Mediterranean, can be bred nearly year-round, and thus may have shorter lactations and more frequent kiddings. Where goat meat is utilized as a fundamental protein source in the diet, the more meat coming along, the better.

Once the doe has given birth, the goatkeeper contrives a routine that extends lactation way beyond the normal nursing period of wild goats. The kid is reared by suckling the dam or by hand-feeding the dam's milk in a bottle or a pan long enough to give the kid a proper growth foundation. The kid is weaned from milk at between six and twelve weeks of age, depending upon the breeder's philosophy and inclinations, but the dam goes on milking. She lactates until approximately sixty days before her next birthing.

The extension of the normal six- or eight-week feral lactation is achieved by regular milkings by the goatkeeper. This demand keeps the doe producing, and the production level is kept up by feeding her a diet that assures her maintenance plus production food intake. The milking is usually done twice a day, morning and night, at twelve-hour intervals. Regular, conscientious milking is what keeps the flow coming. Once the doe's kids are no longer there to stimulate flow and to demand food and suck, you, the goatkeeper, have taken their place and role and must go on creating that demand. If you want and need milk, you keep that flow coming.

At the freshening, the goatkeeper must cope with both the birth and care of the young kids and a doe coming into milk; it is a critical time that requires attentiveness and having your wits about you. Fortunately, milk goats need little human assistance in most cases and have a relatively easy delivery. At this point the cycle starts all over again with the doe kids. They are reared on milk, weaned to a herbivorous diet, raised to doelings, watched for the onset of regular heats and bred during one of them, cared for through the five-month carrying period, and then freshened to become milk goats.

In dairying, one is dealing with an artificial situation. Instead of species' perpetuation being the reason for breeding, humanity breeds dairy animals to obtain a milk supply for human consumption over an extended period. Without getting into an ethical debate about whether or not this is exploitation, it is still a good idea to keep the contrived nature of the scene somewhere at the back of one's mind. Many of the difficulties folks meet up with when they start keeping milk animals arise from the inherent artificialities of the situation. At some points in the cycle, the goatkeeper needs to imitate the structures of the undomesticated goat, and at other times it is necessary to find regimens that differ but simulate the instinctual patterns. If you want milk, healthy dams and kids, and contented goats, aim for a balance between the instinctual and contrived modes.

CHAPTER 4

MILKING

Milking is the functional focus of goatkeeping and one of its primary pleasures. I stress milking because I think milk yield is the most important consideration for the household keeper. The breeder's focus lies on stock improvement; this of course entails milk productivity as a major goal but other considerations, such as conformity to breed-type and showing standards, receive substantial emphasis. For the household keeper, though, a goat's "milk potential" is a matter of what finds its way into the milk pail. The rhapsody about the pleasures of milking will be brief because it is one of those acts you must experience for yourself.

How to Do It

There are numerous verbal explanations of how to milk a goat. You get yourself comfortable and the doe stationary. If the doe has been milked before and has had good relations with the folks who have previously milked her, she'll stand for you as long as she is rewarded.

The reward is the grain ration. I have heard of people who successfully milk their goats without feeding a grain ration during milking — I just have never met any such goatkeepers in the flesh. It is quite possible, I think, but probably works only for a very experienced person milking a very good-tempered, well-trained goat.

I always set the traditional reward before the eyes and eager lips of my does, and lock them into a milking stand with a crossbar. I then sit facing the udder and put my hands in place. I feel around the udder gently and check for any unusual signs: extraordinary warmth, lumps, cuts. As one becomes better acquainted with one's does, this initial touching will become more confident and more and more sensitive. I then dip each teat in a warm iodine- or chlorine-based washing solution. I wipe each teat dry with a disposable paper towel. (I resisted using disposables until the washing and sterilizing of the goat house laundry — wiping towels and cheesecloth filters — got out of hand. Once we had three milkers we seemed to be consuming as much water, soap, pumping energy, and so on, as using throwaway products might involve in terms of consumed resources.)

After washing and wiping, I milk one or two squirts from each teat into a separate little pan called a strip cup, and then check this initial milk for oddities: clots, flakiness, bloody streaks, the possible signs of mastitis-type infections. Washing has cleansed the udder area of particles that may be carrying spoilage bacteria, and these initial squirts clear the teat canal itself of foreign particles or unfavorable microorganisms. This sounds like a good bit of trouble, and it is, but this routine is fundamental to good-quality, good-tasting, longer-lasting milk. If you want to drink your home-produced goat milk raw, you'll appreciate these precautions.

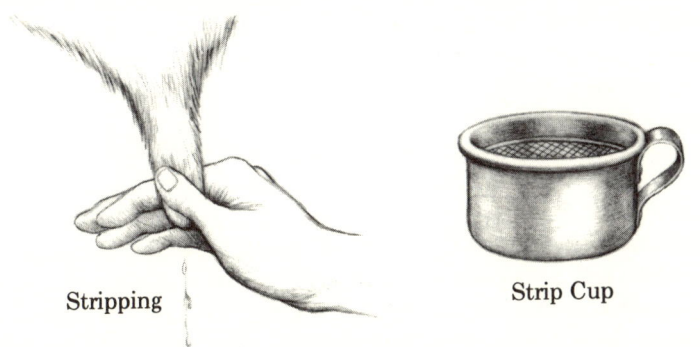

Stripping

Strip Cup

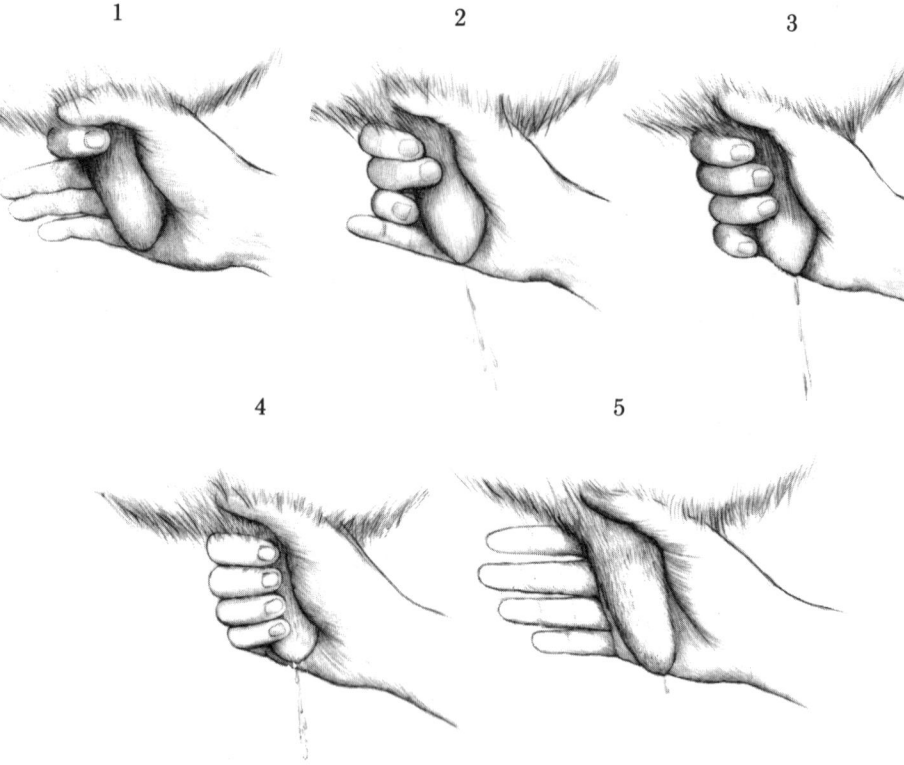

The essential milking technique consists of making a ring around each teat with your first few fingers and thumb, and then pressing downward with your last fingers. You can release the circle and use four fingers to press the milk down and out of the teat, if necessary. Simply reform the circle and repeat the pressure. A slight break between downward-pushing pressures allows the teat to fill up again. Eventually work toward a rapid, even motion, alternating pressure on each teat, or pressing them both at once and pausing slightly for the teats to refill. Avoid pulling down on the teats or the udder. It is tempting to pull down hard to get a good initial flow going or to strip out the very last drops, but it strains the udder musculature. Not having their udder muscles punished or stressed will enable your does to have the longest possible productive life. If you wish to empty the udder completely, hold the teats and gently push upwards against the bag; then continue milking out. Pushing upwards may seem contrary to your judgment, but watch young kids nurse. They hit the bag upwards to stimulate the flow and the more aggressive kids punch hard. The reason many breeders separate kids and dams is to avoid the continuous, aggressive punching the dam is subjected to by some of the kids. A gentle suggestion of this upward thrust usually serves to bring more milk flowing down.

All of this assumes that you are dealing with an easily managed doe with well-formed teats and a reasonable amount of milk. My initial situation was so different from this that I wonder how I ever milked Mackenzie at all. Her teats are not well delineated or set off from the rest of the udder; I never have found any special place to put my thumb and index finger. Her bag is pendulous to an extreme and lopsided besides, the far teat hanging two inches lower than the nearer teat. To compound the problem, her bag was incredibly full every twelve hours and she inevitably ate her way through all her grain ration before I milked her out. So, don't give up hope, however extreme your first experiences.

Getting Practice

Before practicing on your own goat, I recommend some prior preparation. Learn to milk a friend's goat. Seeing it done by a gentle, experienced keeper was worth all the diagrams, verbal descriptions, and photographs I'd ever seen. Start at the tail end of the lactation period when the milk flow is lessening. It is extremely frustrating to make first attempts on goats that are carrying four to six pounds to every encounter. It takes time and practice to develop your hands and accustom them to this new kind of work. Get some heavy practice sessions under your belt, especially if you are currently raising doe kids and plan to be milking them as first fresheners. First fresheners can be very skittish — their udders taut, and their teat openings tight.

Any serious keeper or breeder will have a genuine concern in helping you learn to milk, because the majority of goatkeeping people like to spread the faith and see that newcomers treat their animals properly and tenderly. That objective usually involves showing persons how to do it — the correct, effective, nonpulling way. Such keepers don't hold with the old saw that strangers in the barn unnerve goats, make them difficult at milking time, and lower the yield. As for more negative-minded goat owners who are adamant about barring you from the barn at milking time, pass them by. You are bound to find a more interested and cooperative person down the road if you keep looking.

Recent homesteading literature is big on advising people to milk balloons or filled rubber gloves. This type of practice didn't help me and I discovered eventually that it hardly related to the real thing on any level. It might make a good children's party game or give you a deep laugh some cold, long January evening. It might also stand you in good stead if you do come across a doe with teats shaped like the fingers of a latex glove.

Milking Routine

The best milking schedule calls for milking at twelve-hour intervals. Goats' productivity depends upon their feeding, managment, security, and expectations, as well as upon their breed and congenital capacities. If you are regular in your ministrations and they feel they can depend on you, they put out for you. This is a generalization, but it is backed up by broader dairying experience. Keep to a repetitive, equal-interval milking routine. You milk in the morning and again at night. The specific hour is arbitrarily set; you, the keeper, decide. If you farm and have other livestock and chores to attend to, it will probably be most convenient to do the milking early in the morning and then again around dark or early evening when you are closing up your chickens or turning on their lights. If you rise for a full-time job away from your homeplace and leave at 8:30 or 9:00 A.M., you might work in the milkings at 7:00 A.M. and 7:00 P.M. I milk at 9:00 A.M. and 9:00 P.M., because my early morning is crammed with activities. This time arrangement also works out well if someone has to spell me when I am away for the day. My situation has currently evolved into a two-person labor input and the established milking schedule fits the second milker's proclivities well. The grain and hay feeding, and watering are done at milking times. When the weather is bad, water and hay are also provided at midday.

Where you milk, is your own choice. Outside milking, when feasible, is the choice of some small-scale keepers. The sun and the totally ventilated, unenclosed situation both contribute to a healthy and sanitary milking environment. Whether you use a milkstand or not is up to you. For inexperienced persons and new goatkeepers dealing with old goats who are set in their ways, I believe the stand saves labor, encourages new habits appropriate for the new situation, and simplifies life for both people and animals. I feel this is particularly true if you are keeping goats as a one-person endeavor. My old Mackenzie was accustomed to a milking routine that was based upon two persons being present for every milking; one person milked her and the other held her by the horns to keep her still. I would never have been able to milk her by myself if I had not had the psychic and physical support of the stand. With her changed habits, she is a shining example of dairy goat behavior on the milk stand, and I have been able to train her daughters and other young does to pleasant behavior at milking times. The grain reward and the example of a wise, old, she-goat have made for conditioned responses in the classic sense; the does eagerly hop up to the stand to be fed and milked. Knowing that your milkers will behave in this way makes a huge difference to you when milking time comes in sub-zero weather.

Cleanliness

Debates can go on and on about where to build or place the milk stand. All literature oriented toward serious goat dairying and commercial milk production insists upon a separate milk parlor. Milking the goats in a room separate from their stalling areas helps to avoid high coliform bacteria counts in the milk and insures a milking area that is relatively free of loose floating hairs, dried bits of hay, manure particles, and any other chance vehicles that the bacteria hitch a ride on. An isolated area is more easily scrubbed down and sanitized. The more goats you keep, the more necessary a separate milking parlor may become for you. Much depends on whether you plan to sell milk and whether you have any negative experiences once the numbers in your herd dramatically increase.

We have all enjoyed observably better health since we've switched to raw goat milk. Our home-produced milk keeps long and well and has no goaty odor or taste. We do not have a separate milk parlor. I milk each doe on a stand that is six feet from the nearest stall. After milking I scrub the stand down with an antibacterial solution. I mentioned earlier that I dip and wash down the does' udders before milking. I now also give each milker a teat dip after milking, using a very dilute solution of ordinary chlorine bleach. We started this after noticing the large teat openings of one of our first fresheners. Large teat openings can provide entry for pathogenic bacteria. By adhering to careful sanitary practices such as these, you probably won't find it necessary to build a separate milk parlor, especially if your milk is for family consumption.

To achieve peace of mind regarding TB and brucellosis, the most serious milk-transmitted diseases, have your milking goats tested. These tests are routine and inexpensive, and any livestock veterinarian will do them. If a goat of unknown origins comes to your barn, have her tested — for your own safety as well as that of your animals. Our local vet says that our area has been brucellosis-free for twenty-five years, and that he has never run into a goat carrying brucellosis or bovine TB, but we still test.

If you're still worried about milk-borne disease, you may also choose to pasteurize your milk. I believe that raw milk is better for me, so I give substantial attention to sanitation. The barn routines mentioned above and dutiful scrubbing and scalding of milk utensils have contributed toward wonderful milk we have all enjoyed and that visitors could not discern as "weird," only "tasty."

If someone politely comments that the milk is refreshingly "wild, strong, gamey-tasting though exciting in its way," it is probably time to review your barn and milking procedures. Politeness accounts for these euphemistic ways of saying "unpleasant" and "off-tasting." They are signals that your desired quality improvement has not been achieved or that the standard has slipped.

Once the initial insecurities and trepidations you and your first milker both experience ease, you will probably find that milking a goat is just plain pleasurable. She is a small, fastidious, affectionate, clever creature. She is not as placid as most cows are reputed to be, but the doe is cooperative in an intelligent, responsive manner when treated with respect, regularity, and human affection. It is a physical joy to feel that warm bag, squeeze the teats, see and hear the milk pour forth, and grow connected to the whole process of nourishment. There are hundreds of small gurglings and all kinds of inner rumbles that your ear will catch as your head leans against a doe's side. Milking is a small sacrament of the new life, which is one and the same as the old life. Thinking on the transformation of scraggly herbage, you witness and take part in the making of a warm, new, foamy, yet solid, life-sustaining substance — milk!

CHAPTER 5

BREEDING

Breeding is mating. It influences many facets of goatkeeping. Cogenital capacity determines to a large degree your doe's lifetime productive capacity. Good genetic inheritance combines with good care and management to give the best milkers and the most milk in the pail. When improved dairy goat breeding was begun in the United States, breeders found that the first-generation doe bred out of a scrubby, native, wild-type dam and an imported, selected Swiss stock sire showed enormous increases in milk yield and in duration of the lactation period. Successive breeding of the improved does and their female progeny to imported, purebred Saanen and Toggenburg sires yielded further improvements. You must take a long, almost unbelieving look at the 1930s research station photographs of specimens of the native goat stock even to be able to recognize the subjects as milk goats. At a quick glance, the goats look like horned coyotes; legs are short, udders tiny, necks short, and jaws small and undeveloped.[3] There is little barrel, or lung capacity. The positive evolution, visible and since recorded by contemporary testing and statistical evidence, is undeniable. It furnishes the foundation argument for breeding-up.

"Breeding-up" refers to the practice of mating does of lesser or unknown origins to proven-out purebred sires. It attempts, over several generations, to produce does that give more milk for more sustained lengths of time. A 1967 Cornell dairy goat bulletin remarks that three-month lactations are not uncommon,[4] but the three-month lactation has become increasingly rare as of this writing. Conscious breeding-up, or upgrading, over the last thirty years has produced recorded grades with ten-month lactations, as well as enormous productivity increases among registered purebreds, crossbred animals, and new breeds.[5]

A Fundamental Experience

The breeding of Eleanor R. Toggenburg was our fundamental goat-breeding experience. Eleanor was an unregistered, stately, Toggenburg-resembling, chocolate-colored, maiden goat of mixed and unrecorded ancestry. She was past two years of age and had never been bred. We were boarding her, and were to keep a kid out of her as the board fee. We took the responsibility for breeding her and made the choice of sire, electing to make a journey of sixty-five miles (each way) to breed her to a French Alpine buck — the breed I was most interested in — totally unrelated to any stock close to us. Another reason for choosing the breeder was that he was one of the few who did not object to Eleanor's having horns. I had reseached the stud-service resources in advance in anticipation of the great day.

Eleanor, however, never seemed to be very certain about when her great time was upon her, or if she knew, she sure didn't show it. She had a skittish temperament normally, seeming at times more wild antelope than dairy goat. Her sexual life conformed to this fickle pattern, and was equally difficult to interpret. She never did have observable, three-day heats, twenty-one days apart. Her tail would wag from side to side and her vulva pinken considerably, but for hours only, and at odd intervals over a month. She wasn't in milk so she couldn't show a decrease in milk yield. Her appetite never observably decreased. She never became vocal about her sexual needs. Finally, one day, when we thought we had seen sufficient tail wagging and genital reddishness, we urged her into the microbus and drove with her — frightened, lonely, and still silent — to the breeder's. Well, she was in no way interested. The breeder had four different purebred bucks and a barn properly saturated with buck scent. Eleanor was unimpressed. And were we embarrassed! The breeder comforted us with stories of other quirky goats and beginning owners. After awhile, we decided to board Eleanor at the breeder's until her sexual needs could be fulfilled.

Many times that night we asked ourselves why we hauled a plain old scrub so far to get her serviced. We came up with the same answers that had set us on the road in the first place; only time could provide us with other justification.

Breeding Up

Our doe, Mackenzie, and our boarder, Eleanor, were half-sisters fathered by the same buck. Many of the other goats in our area were sired by that buck or came out of a dam sired by that buck. Mackenzie had even been bred to her own father and had borne Bela, the doe kid I purchased at the Library Fair. Human horrors over incest aside, inbreeding is a common practice in sophisticated animal breeding. Indeed, careful inbreeding is the foundation of many improved herds. Matings between the closest of animals — fathers and daughters or brothers and sisters — have traditionally been used by experienced breeders to develop breed-type and to perpetuate specific characteristics, but it is never the recommended practice for small-scale or novice keepers. Close inbreeding can perpetuate weaknesses as well as strengths, and since goats do not have huge litters, culling two out of three kids for congenital weaknesses can mean a great loss for the small keeper. Most small-scale keepers simply can't afford to cull rigorously, and can't afford to carry poor does to maturity. So the closest matings are not advisable.

A variant of strict inbreeding, linebreeding, depends upon using related animals but not the most closely related. The purpose is to maintain a high degree of relationship to proven-out, superior animals. Linebreeding carries a lesser chance of perpetuating weak traits and is more generally recommended as a breeding system for the small keeper. But we decided to import some new blood.

What we decided to do is called "outcrossing," the mating of totally unrelated sires and dams. This is far from the usual choice in breeding circles. Theoretically, stock improvement comes very slowly when animals of widely differing gene pools are mated. However, we decided that too many animals in our vicinity shared the same unselected, mediocre genetic inheritance. We also realized that it would be years before we could judge the results of Mackenzie's having been bred back to her own sire. We had to decide what to do for Eleanor before any data on Bela would be in. We either had to breed Eleanor back to her own sire or take her out of the area to a different buck — breeding her up in the process. Over several generations of breeding-up, the genetic inheritance becomes increasingly more that of the purebred sire. Our purpose was to import into our mini-flock the beginnings of an improved line that had no relation to the goats in our immediate locale and would be of a breed of our own choosing. This method of improvement is slow. It will be generations and, we figure, three to five years before we have good foundation stock of the breed we are interested in. Eleanor's daughter, Meg, is a half-grade French Alpine; Meg's daughter a three-quarter-grade French Alpine, and a daughter of hers would be a seven-eighths or American Alpine.

Every goatkeeping authority advises breeding-up. The broad purposes are to improve milk production and the reproductive capacity of one's stock, to maintain true breed-type, and to prevent the appearance of regressive or deteriorated individual animals. Underlying all these abstract purposes is the blunt fact that kids bred out of registered, purebred sires can be recorded and registered, and therefore sell for appreciably more money than kids bred out of a couple of unknown goats. The theory is that a kid from a registered, recorded line has more traceable potential as a milk animal than kids of unknown origins. However, the registry rules read as follows:

> *Rule 14. Where one parent is a known registered purebred, the female offspring may be recorded as a ½ grade of that breed. If such offspring is then mated to a purebred male of the same breed, the resulting female offspring may be recorded as ¾ grade of that breed. These ¾ grade females when mated to a purebred sire of the same breed will produce kids that are ⅞ pure and the males and females of this group are eligible for entry into the "American" section of the register providing they meet the breed-type requirements. (Rules for Registration and Recordation, ADGA)*

Consequently, there are many registered kids on the market that aren't from very "milky" backgrounds because *any* progeny from *any* purebred may be registered at the breeder's discretion. What is happening today, a time of burgeoning interest in goats, is that many people rush to breed-up their unregistered or recorded grades simply because they can charge more for the registerable kids and they need the income from the sale of kids to sustain their goatkeeping operations. This is really nothing but paper breeding-up and paper purchasing, and so it accounts for the large number of goats with papers but no milk productivity.

Choosing the Sire

A distinction should be carried in mind between any available, papered, purebred or American sire, and the positive, proven-out buck. A buck that has productive, tested and recorded female offspring and near relatives (sisters, dam, granddam) that are good producers is a proven-out buck. If you're just breeding your does to any available papered buck simply because the kids will bring in more income, you're not necessarily breeding kids that are improved animals. To justify the added tariff you charge for the kids, try to breed the does to proven-out, papered bucks.

Outstanding bucks are those who are not only proven-out, but who have sired female offspring that demonstrate better observable and recordable milk productivity than the dams that produced them and better than offspring of the dams by other sires. Breeding fees for the services of outstanding proven-out bucks will probably be high. When your herd is evolved enough so that you consider it a high-producing herd, you may want to invest in the services of a truly outstanding buck to attempt breeding an even higher yielding doe. For someone starting with unregistered does this is a distant prospect.

Breeders all say that the buck is half the herd. In the case of the unregistered, scrub herd, the bucks chosen at breeding season represent *more* than half of the future herd. They contribute, over several generations of breeding-up, the dominant share of the recognized dairy character in the gene pool. The traits of the original scrub doe recede far into past history within three generations. In the long run, careful breeding-up yields does of good type and solid performance, and more milk in the pails.

Mating

The heat period, or estrus, is the only time during which a doe may be bred. After the pituitary gland has been stimulated by seasonal factors such as diminishing daylight and temperature decreases, it secretes the follicle-stimulating hormone (FSH), which stimulates ovarian activity, which in turn causes the secretion of estrogen. This hormone can produce many behavioral changes. The doe loses her appetite but finds her voice. Her usual, bright, clever, attentive personality undergoes a radical transformation toward moony, bleating, erratic, uncustomary behavior. She seems to forget the route to and purpose of the milkstand, and wanders instead from her stall toward the barn door, occasionally poking her nose out into the night. Grain suddenly has little or no appeal for her and milk yield drops.

Some does are very cooperative and exhibit the blatant physical signs of heat: rapid back-and-forth tail-wagging; a swollen, pink, red, or mucousy vulva; frequent urination; bleating; and nervous, uneasy activity including the mounting and riding of her companions. Textbook signs like these are most reliable, almost a sure thing, but don't count on any of them appearing. Some goats follow the pattern of twenty-one day intervals with heats of approximately three days; others follow their own signs and interval calendars. There are authorities who recommend breeding on the second or third day of a heat, but I bring my does to buck as soon as they show definite signs of a heat. The best artificial insemination results have been obtained when the breeding was done within the earliest stage of the heat, and many old-timers advise striking when the iron is hot — within the first twenty-four hours when the doe is sure of what she wants. The best diagnostician of the doe's state of need is a real live buck; his degree of interest will tell you right off.

A good, vigorous buck will let you know in a short time what his analysis of her biological condition is. If she stands for him and he recognizes her as being receptive, he mounts her in a flash, penetrates the doe with his penis, and ejaculates semen faster than it takes to write all this out. And is it all brief! Don't carry any anthropomorphic or romantic notions of sexual enjoyment to service. Return runs are very common. Not everyone who keeps a breed buck controls the scene tightly and brings the doe to the buck at an observed time under restraint — such as being held by collars or tethers. Some people just allow the doe and buck to run together. Return trips are more common under this looser, less controlled arrangement.

Don't feel too bad if you find upon arrival at the breeder's barn that your doe has suddenly gone out of heat, or that she will not stand for the buck, or even butts him away. The buck may show no interest in her either. His perception of her state is usually accurate. You may have noticed her signs only late in the heat. With delays and transportation difficulties to deal with, arriving at the buck barn with an uncompliant doe is not out of the ordinary. If this is what happened, note the date of the first observable signs of heat on a calendar, and then mark the day of the month eighteen days from the observed signs. Watch the doe closely from the sixteenth day on. If, after another rushed and breathless try, you or she somehow misses again, it may be better to work out an arrangement with the breeder to board your doe for a few days in advance of the heat. Young does in their first season and older, unbred maidens, seem to cause the most problems.

In any event, watch the doe closely during service. If the buck mounted her, and there seemed to be an apparent service, semen ejaculation, and consequent calming down of the doe, you have to observe her carefully all the next month. If there are no recurrent heat signs during this observation period, your doe has been bred — she's in-kid. Write down the date of the service if you have no written stud memo as a record, and count 150 days (this is an average) ahead. Milk is on the horizon in less than five months, for at least eighteen to twenty-one days have passed and your doe has not come back into heat.

Sometimes around the dates of what would have been another estrus period, you may see some tail-wagging, bleating, mounting, or pinkening at the vulva. This is confusing, and may put you back on the road to the breeder, but once there, you may find the buck has no interest. Eleanor, our original breeding instructress, showed such signs several times after her mating and turned out to be in-kid after all. On the whole, one just has to stay alert for any signs of heat behavior after service and use one's judgment and increasing experience about whether the signs are consistent and extended enough to signify a true return of heat.

To add a little more doubt to the matter, I have to say that you can't rely on "seeing something" as an indicator of whether or not your doe is pregnant. The growth of the young is not rapid, or visible enough for you to be able to count on seeing whether the doe is in-kid in the first eight weeks after the service. Young does and smaller-framed does do show the kids they are carrying earlier than mature or large-framed animals. Some does of varying sizes and ages always from the outset of the pregnancy look like *burritos* carrying packs. Eleanor drove us to distraction with her brief false heats and her lack of visible physical alteration during the gestation period. At six weeks prior to freshening, even our vet was unsure as to whether or not she was in-kid; but bred she was.

The breeding season runs from September to March in my region. Actually, the doe has an innate, instinctual calendar that is affected by temperature and light changes. The seat of this calendar is her pituitary gland. As the days shorten, light decreases and temperatures drop. The pituitary reacts to these stimuli, particularly to the light changes, and secretes a follicle-stimulating hormone that, in turn, stimulates the production of estrogen that triggers the heat period. Most, but not all, dairy goats will only come into heat during this fall period, what everyone calls "the season." However, Nubians and other goats of non-Swiss type may come into heat at other times of the year. In any event, the bulk of breedings in my region take place in the fall and early winter. Many small-scale keepers do not want kids arriving in the deep winter months because heat lamps or special arrangements may be necessary and because it is sometimes difficult to sell the kids in the dead of winter. Visible heat signs, durations, and intervals vary greatly among individual goats, but the keeper will have numerous opportunities to breed does in any given season. Once you are experienced with your particular animals and the general behavior of the herd during heats and have decided where to breed them, you can plot a breeding schedule, determining which does to breed early, which does to breed late, and whether to stagger the breedings or not. These options should be decided upon in relation to your individual needs.

Keep notes. Record where, when, and to which sire you've bred a doe. Obtain an ADGA or other registry association stud service memo (the original goes to the owner of the doe and the duplicate to the owner of the buck) when breeding does to registered bucks. Even if you don't plan to register the offspring, you should keep notes. They are your source for accurate dates from which to calculate the freshening or due date. They also give you a basis for detecting irregularities, problems, or breeding complications such as infertility and they provide background information so that does are not bred back to their sires or brothers unintentionally.

Arranging Stud Service

Locating bucks for stud service is not difficult, but, contrary to popular belief, you don't just smell them out. Stud service resources should be checked out in advance of the season and contact made with local breeders who keep bucks. You can't count on every breeder being available just when your doe comes into heat. If you've given no advance notice, you can't count on the breeder's availability or the availability of boarding facilities. The best guide to information on available stud service is the up-to-date members' directory published by the ADGA. The directory lists members, addresses, the type of goats each breeder keeps and, if they have bucks at stud, the breed of these. Your local goat club, the state dairy goat breeder's associations, and specialized breed clubs also have information on the bucks available for stud service. These information resources are for registered purebred and American-designated bucks. Of course, there are many unregistered sires kept by small-scale keepers. Before bringing your does to these bucks, check out their female offspring, and research possible relationship factors in your doe's background; the buck may be her own sire. If so, you may want to try a different buck. If the buck in question is the only available stud, breed the doe back to her sire, but remember that chances are good that inbreeding will perpetuate weaknesses as well as good traits, and that such close in-breeding is recommended only for experienced, genetically informed, well-financed professional breeders.

If you are making many return trips and seem unable to get the does bred, there are several factors that may be considered. The buck may be too young (or too old) to be fertile; there may be hormonal problems in the doe or buck; protein, mineral, or vitamin deficiencies may exist in the buck or doe; obesity or disease may be a factor. Lack of fertility is probably the most frequent problem; the others are uncommon. If you are a beginner, regard youself as the most likely cause of breeding misses. These often come about through inexperience, inconsistent observation, and inadequate recording of heat periods. If after tightening up your observation and calendaring you still haven't bred the does successfully, try a different buck. If you have serious, persistent difficulties, consult your vet and chapter ten in this book. As compared to other domestic productive livestock, the goat has a relatively easy and straightforward time of it when it comes to reproduction, and the keeper of two or three does will only occasionally encounter grave breeding difficulties.

If anything, goats breed so easily that early or premature breeding of young does is common; much of it is accidental, but some is intentional. Ten months or eighty-five pounds of bodyweight are the usual guidelines for a first breeding. Many keepers wait until their young does are one year or eighteen months of age before breeding them. If the doe is bred too early, competition may arise between the doe's own maintenance and growth needs and the demands of the fetus and early milk production. The doe herself grows little in her body frame or digestive organs during the last eight weeks of pregnancy and the first months of lactation. It is possible that early-bred does may be permanently stunted in frame development and rumen capacity, which, in turn, stunts long-run productive capacity. However, this is not inevitable. If the doe kids healthily after an early pregnancy, is fed adequately in the latter lactating months, and is not bred and run through a long lactation the next year, she may mature more completely. You may also cut off her first lactation at four or five months so that food input and energy will go towards further growth and rumen development rather than milk production. In conscientious hands early breeding difficulties may be overcome, but it is important to decide on a management compensation arrangement for the following year that will allow the doe to mature without production or pregnancy stresses competing with her fuller growth.

Breeds

So far, I have avoided any discussion of the advantages of owning animals from one of the particular, recognized dairy goat breeds. This has been deliberate on my part. Most goat literature goes into great detail on the subject of the breeds before any material on the process and purpose of mating is presented. I have reversed the more usual pattern to give physiological and empirical aspects of breeding more emphasis. I feel that deciding upon a particular breed boils down to what female stock is most immediately available and to geographic factors, such as the specific breed of the nearest good buck. The day is past when one breed could be recommended without qualifications over another for quantity of milk, cream content, temperament, or disposition. Recorded stock improvement is evolving rapidly; new breeds have been created, and new productivity heights achieved. What this means in terms of deciding on a breed is elaborated somewhat in the following pages. It's a time of transition, with much work being done on improvement of productivity within each breed and in creating the new La Mancha breed. Because so many more people are keeping goats and recording their productivity, it is a seed time for future movements in goat dairying.

All contemporary dairy goats are seen as amazingly productive when their statistics are compared to those of goats of a generation ago. Swiss-type goats have always yielded the most milk. The specific Swiss breeds are the Saanen and the Toggenburg, both of Swiss mountain origins and both produced by selective breeding practices.

The Saanen is a white-haired (or cream colored), pink-skinned breed of medium to large frame. Wattles and dark freckling on the bag and skin may appear on Saanens. Their udders are large, well shaped, and well hung. Saanens are often big goats, and are the backbone utility milk does for many people, since good Saanens can be phenomenal producers and their milk has a good average butterfat content. A British-bred Saanen, Melpas Melba, had a recorded milk yield of 6,661 pounds in 365 days.

Toggenburgs are equally sturdy, vigorous, and productive goats, though often not as large framed as Saanens. The breed is characterized by white blazes running down the sides of their faces, and light or white markings on the lower legs, around the tail, and near the wattles. The body background color may be any shade of brown from

This Saanen is GCH Wil-Lea Kalen, a star milker, classified and rated excellent at 90.

This Toggenburg is GCH Diamond Sunshine Zesta, a four-star milker, classified and rated excellent at 92.

deep chocolate to pale beige or fawn. There are long- and short-haired Toggenburgs; short-haired does are often preferred for reasons of dairy sanitation. In 1957, David Mackenzie noted that the Toggenburgs of the British Isles yielded below 1,600 pounds of milk annually and that the milk contained only 3 percent butter fat.[6] This situation has changed, especially in America. In the United States a Toggenburg holds the all-time, all-breed milk production record of 5,750 pounds in 305 days. Concurrent increases in butterfat content have also been achieved by Toggenburgs.[7]

The French Alpine is a large animal characterized by a great variety of colors and markings. Certain patterns are historically recognized and designated, among them:

   cou blanc — white neck, black hindquarters
   cou clair — tan neck, black hindquarters
   cou noir — black neck, light hindquarters
   chamoisé(e) — light brown or chamois colored
   sundgau — black with white underbody and white facial
      markings
   pied — painted or black and white combinations
The French Alpines also appear in a large variety of unclassified coats: browns, whites, cinnamon. The purebred French Alpines in the United States trace their ancestry to a 1922 direct importation of French stock.[8]

These three mountain-derived breeds share closely related conformation and characteristics. Their faces are straight or dished; their ears are upright; their carriages similar. They have formed the quantitative and qualitative production backbone of many American herds, large and small, since the original introductions of European purebred stock to the United States in the first quarter of this century.

The Nubian breed in America derives from the Anglo-Nubian of the British Isles bred there by crossing European and Eastern stock. Nubians are characterized by a Roman nose, unmistakably pendant or floppy ears, and a wide and beautiful assortment of coat markings and colors. Their milk has always been distinctive for its high butterfat content. Recently a Nubian doe surpassed the 1960 all-breed butterfat production record held by a Toggenburg with a recorded 224 pounds butterfat content in a 305-day lactation, and also set a new Nubian breed quantitative record of 4,420 pounds for the lactation period. It used to be a truism that Nubians gave less quantitatively, but more qualitatively because of the higher cream content of their milk. British breeders first took advantage of the fact that goats from

This French Alpine is GCH Diamond Sunshine Stella, a three-star milker, classified and rated excellent at 90.

This Nubian is GCH Dollie Sunshine Cindy.

warm climates gave rich milk while those of European stock gave high yields. In recent years American breeders have further improved herds by selective breeding and have generated Nubians that give high yields of high-butterfat-content milk.

Another sign of transition during this dynamic period is the arrival on the scene of a fifth major breed, the American La Mancha. The La Mancha is a newly evolved category developed by crossing a short-eared Spanish-derived stock with leading American purebred stock. Colors and markings vary widely. The distinguishing traits are the absent or short ("gopher" or "cookie") external ear; short, glossy hair; and straight faces. Neither the quantity nor the butterfat content of La Mancha's milk is as high as that of older, more established breeds, but there is much attention being given the new arrival.

These, then, are the major breeds along with some of the records made by extraordinary animals of the currently leading herds. Do remember that record breakers and record setters are not average animals of their breeds.

If you are interested in higher milk yield over a long-run period, breed your present does to the best proven-out sires to which you can reasonably transport them, and for which you can afford the fees. It is more efficient for the small keeper to maintain two or three good-producing does than five mediocre milkers. Keep the female progeny out of your doe's best matings. Rear them to milking age, and you will have replacements for your current does that, should all go well, will be even better producers than their dams. If you are selling all the kids, you owe it to yourself, your buyers, and the evolution of goat dairying in general, to try to breed kids that are as good as or better than their dams. Then, you can in good conscience get good prices and contribute to general stock improvement at the same time. In effect, I'm suggesting that you have a stake in future productivity that should be as great as your concern for higher income from the sale of kids. Productivity is high today because of what we've inherited from breeders over the last thirty or forty years. All of us recent arrivals in goatkeeping have a debt to future goat dairying. We must see to it that the recent upsurge in goat population that we have contributed to heavily maintains a good level of productivity and doesn't just populate the countryside with caprine pets. You should take on a caring attitude in this respect. But do not take on the responsibilities, stance, and concerns of the professional breeder unless you are prepared with backup money, assets, land, and experience.

CHAPTER 6

BIRTHING
AND
KID-REARING

At the core of any solid dairy enterprise, even the smallest, is the handling of freshenings and the raising of young stock. The care and rearing of the young provides replacement stock, substantial income, and gratifying entertainment.

The milk animal comes into milk when she delivers her young. The easiest and most rewarding deliveries are those that are prepared for well in advance. Consider the in-kid doe's diet first. It is vital that she be adequately nourished but that she not be drastically overfed with concentrates. This may result in a protein overdose and delivery problems. This translates as "go easy on the grain." I have had to resist those inner voices that whisper, "give her a few more cupfuls — she is eating for two, or three, or four." Adjust the grain allowance in relation to the doe's age, her past kidding and production records, the quality of the hay being fed or the available pasturage, and her general size and maintenance requirements. Additionally, it is wise to substitute bran for part of the in-kid doe's grain ration during the last four to six weeks of the gestation period. This keeps all her digestive

51

channels open and functioning regularly. Also, since I incline toward folk remedies, even for goats, I give my pregnant does a few table-spoonfuls of apple cider vinegar in their grain ration. If your does resist eating vinegar-soaked grain add some diluted unsulphured molasses. I used to feed vinegar with the grain only for the last sixty days, but now I use it on the ration throughout the entire gestation period, starting with one tablespoon per feeding at the beginning and working up to three per feeding in the last weeks. I prepare my own vinegar by adding a homemade apple cider "mother" culture to a gallon of unpasteurized, fresh cider, and keeping it warm till it begins to turn. I first learned of this practice in Dr. Jarvis' *Folk Medicine* where he writes of disappearing bovine mastitis in herds fed cider vinegar. The practice confounds most livestock veterinarians and local extension agents, but since my goats do not have mastitis or other inflammatory problems during their dry months or at freshening, I stick with my few ounces of vinegar prevention.

Drying Off the Does

Goats and cows are dried off for two months prior to birthing so that they do not get overly stressed by the double demand of producing milk and bearing the fast-developing young. For the strongest kids and best eventual milk yields it is good to follow this practice. I found, though, that drying goats off turned out to be more difficult than the literature indicated it would be. Different methods abound and I've heard much contradictory advice. The most useful, direct, and simple technique that I have used is called the Kraftborner method. Approximately eight weeks before the kidding date, cut down on the number of times you milk daily by milking only once a day for three days. Then milk once a day on alternate days for three days. Then simply stop milking your doe even though her udder still fills and swells. Continue to feed her at the milkstand, checking her twice daily for any unusual udder signs. Continue to cleanse and teat-dip her udder as usual. By ending the demand, you signal her to scale down her production and then to end it. The milk will be gradually reabsorbed into her system and she will stop producing. This will not be a great stress, though her bag will look very full and tight.

In my observations, the reabsorption takes from six to ten days. Its end is shown by the doe's bag going loose and flaccid. After ten months of milking, your doe will not be producing peak amounts of milk. Her udder withstood carrying eight to twelve pounds of milk a day for all the early months of the lactation. It has the capacity to carry this much for several days while her body slows production and begins reabsorption. Keep an eye out for unusual heat, swellings, or lumps. These are the first signs of udder inflammation. Bring her to the milk stand as usual, but simply massage, observe, and cleanse the udder while she eats. It is a helpful practice to bring first fresheners to the stand for a similar observation and handling. It will introduce them to the idea of eating their grain on the stand and will accustom them to having their udders handled.

It is possible to run into an occasional heavy producer whose udder doesn't get flaccid after two to three weeks. Such a goat may have to be milked out a few times a week to avoid serious congestion or inflammation. Such a doe may never get totally dried off for several years running. As long as you drastically reduce her milk productivity over the last sixty days, she will probably come along well, both at the freshening and in the next lactation period. I have had experience with a very productive doe who would not dry up. I saw her freshen with healthy, large twins two years in a row and return to her normal ten- to twelve-pound productivity after freshening, despite the fact that her udder never got a complete rest. Some keepers incline toward cutting grain rations in half and reducing water intake when they begin to reduce the frequency of milkings. I have done this with some of my known heavier producers, but do not any longer, for it risks having some really hungry does in the barn just at a time when their young are putting on substantial growth.

It is common today to routinely administer antibiotics to dairy animals during their dry period. There is strongly divided opinion among veterinarians as to whether this cow-oriented practice should be adopted by goatkeepers. Goats that show *any abnormalities* of the udder at this time should be checked by a veterinarian. I do not routinely use antibiotics during the dry period and have never had any problems.

Environmental Preparations

At twelve to eight weeks prior to the freshening date, the keeper should begin preparations needed to change the doe's physical environment for the event. A kidding stall should be provided. Keepers should also decide upon methods of rearing and separation (these will be discussed later) so that facilities for the kids, if needed, are constructed by freshening time. This is particularly important for keepers who maintain all their goats in a communal facility. The separation of the dam-to-be from the bustle of the herd helps her to feel more secure — reinforcing her tendency in the wild — and serves to isolate her and the kids(s), while also providing a more sanitary environment. Since I normally keep all my does in separate stalls, I allow the freshening doe to remain in her own stall. This seems advantageous to me because the doe remains in her own mini-environment, isolated from the health problems of other animals, and it makes for minimum fuss, movement, and maintenance. There is less clearing, changing, cleansing, and readying of stalls in general with this arrangement.

Depending upon the weather and time of year, I usually muck out most of the litter and whitewash the stall. This is inadvisable for deep winter kiddings because the bedding provides substantial warmth for the doe and the whitewash often freezes. When I do clean the stall, I leave behind a thin starter layer of decomposing litter before putting in the fresh bedding; I feel this helps maintain the microorganisms in the doe's environment. Instead of regarding all microorganisms as potential pathogens, I see them as part of the natural balance in the doe's surroundings. Whitewashing is a flimsy, surface treatment that has to be constantly renewed, but it is cheap and easily applied (except in freezing weather). It coats stall surfaces with an alkaline layer that discourages the growth of many bacteria that prefer acidic surroundings and it brightens the interior of the stall. An easy whitewash mix is made as follows:

> *Add 5 pounds hydrated lime to 1 gallon water. Mix and let stand overnight.*
>
> *Dissolve 1½ pounds salt in ½ gallon water.*
>
> *Add saltwater to limewater, mix thoroughly and let stand 12 to 24 hours. Stir well before using, and apply thinly and evenly with a wide paintbrush.*

The whitewash will look watery when wet, but will dry to a more opaque coating.

Planning Kid-Rearing, *In Advance*

The keeper can anxiously count the days and hours to freshening, but the actual time is probably going to be selected by the doe. The older, more experienced does will actually choose the time and place with greater insight than the keeper can muster. Many of our does have kidded on that typical low-pressure, slightly moist kind of day that David Mackenzie considers ideal. However, it has been my observation that first fresheners follow no patterns and kid anywhere, anyhow, and any which way their fancy and urgency take them. One yearling first freshener of ours strolled up and down the slope in front of the barn one bright, sunny afternoon, and dropped her two kids from on high without stopping to lie down or let out a single moan. The sun is a fine sanitary aid, so there was not much for me to do but wonder at the doe's choice.

The dam herself is by far the best stirrer of life and general sanitizer. She will lick, massage, and tend her kid as soon as it enters the outside world, and then efficiently deliver a second or third, tending them all within an hour's span. An experienced dam has her progeny delivered, up and about, licked, and fluffed-up within fifteen to forty-five minutes of the first one's presentation. The kids all scamper about, leggy and awkward, and then nuzzle around for their first suck of teat.

And here starts a great debate. Most professional breeders take the kids from the dam before any real suckling takes place. The prime reasons for this are:

— to save wear and tear on the dam's udder;
— to provide equal and measured nourishment to all the kids;
— to avoid difficult weaning situations later on;
— to be able to weigh and record milk yield — you can only estimate yield if the kids are nursing;
— to raise the kids to be people-oriented and tractable.

If you plan to rear the kids by hand away from their dams, it is easiest to take them away before they nurse at all, waiting until the dam has licked and stimulated the kid (immediate removal and toweling off is an emergency measure I employ if the doe is having a problem or triplets). The kids are often not too hungry immediately after birth, so do not become too anxious if the separated kid doesn't accept the bottle or pan the first time you offer it. Immediate separation may be the simplest approach, but it may not be the healthiest. There is much debate about when to remove the kids even among established professionals who are unanimous that the kids should be hand-reared away from the dams.

David Mackenzie and his followers allow the kids to suckle for the first four days of their lives and then separate the kids from the dam. This method assures that the kids get their full share of colostrum, the super-rich first milk that carries the extra antibodies and vitamins the newborns need to get well started. The initial colostrum feedings are absolutely necessary for all kids, regardless of their rearing method. The colostrum gets the kids off properly and is essential to avoid scours and other problems of early life. I have taken kids away from the dams at both stages: during the first few hours after birth before any nursing and after the first four days.

The four-day method may make for sounder starts than bottle- or pan-feeding the kids their colostrum. However, it is not a feasible arrangement unless the keeper has a totally separate facility for the four-day-old kids. If the kids have any hope of milk from their real dams because of hearing their calls or seeing or smelling them, they firmly resist separation by refusing all proffered bottles and pans. The kids will easily become anxious and overwrought in this situation, and so will the keeper. Bottles and pans will be accepted only if there is no further contact with the dams. To prevent contact, keep the kids in a separate building and pasture area. Kids that are immediately removed from their dams on the first day can be returned to the same barn once they have accepted bottles or pans. Acceptance usually occurs the first day. These kids can be back in the barn by the third or fourth day. By then, they have accepted the keepers as their food source and have no interest in the dams. The keeper must provide a separate kid pen for the returnees.

Keepers planning to hand-rear should realize that any hand-rearing method entails being home several times in the day to feed the kids.

Another option is to leave the kids with their dams until six weeks of age. The primary advantages of this method are:

— solid start, good health foundation because of totally available nourishment and mother's care;
— good udder health, in terms of less potential inflammation;
— much less labor for the keeper.

However, kids left on to nurse their dams can be very hard on udder musculature. Bucks in particular often nurse very aggressively. If the doe has more than one kid, it is possible that their growth will be uneven. One kid may be more assertive than another and will get a much larger share of milk. This may be especially true in the case of twins comprised of a buck and a doe. In the event of triplets, I usually take at least one away from the dam. It is very hard on the udder for a doe to nurse three kids, and there is a great chance of a runt among the three if one of the kids is not as pushy as the other two.

Often, kids must be left on their dams because the keepers are unable to feed the kids frequently enough. Many novices do not realize that there should be round-the-clock feeding of kids reared away from the dam and that these feedings should occur on time. Four or five feedings a day are necessary for the first week, including a feeding in the middle of the night. This feeding can be eliminated after the first few days, but four daily feedings are necessary till seven to ten days of age. After that, plan on three feedings a day for many weeks. The stomachs of the young kids are too undeveloped to handle food in large amounts at twelve-hour intervals. Plan to leave the kids on their dams if you have a job away from home.

It is of the greatest importance that the first few days of feedings be colostrum feedings. Some of the major problems of large-scale calf raising have been traced to scanty or no colostrum feedings in the first days of life. Milk out your does and feed this rich substance back to the kids or allow them to get it themselves if they are suckling. Freeze any excess colostrum you milk from the newly fresh dams in their first days. It will serve as backup for kids that are orphaned or whose dams have udder problems or other emergencies. I have not had occasion to use my backup supply for my herd, but have given some to people facing emergencies. Even though the colostrum was from a different dam and therefore had different antibodies, it helped the kids survive. I replenish my stock yearly at the height of spring freshening season and discard unused frozen colostrum from the prior year's batch.

At four weeks of age, kids that have been with their dams can be separated at night and offered grain in the morning, while the other goats are milked. It is to the keeper's economic advantage, and in the best interest of the kid's eventual rumen development, to encourage hay nibbling, outside grazing, and some grain intake at an early age. I feed the kids goat milk until they are twelve weeks of age, but after six weeks they are eating a small amount of grain and are well established on hay or grass.

Many persons feed kids milk replacers because they need the dam's milk for household use or to sell. I am old-fashioned on this score and have had little truck with milk replacers, most of which are formulated for calves and are heavily chemicalized with antibiotics and additives. At best they are a highly processed foodstuff. I would, however, use replacers when the need is imperative, to save the life of an orphaned kid or free milk for your own use. Make sure to use a replacer with a high fat content. It may be a poor long-run decision to raise kids on replacers just for the short-run economic advantage of having milk to sell. Again, this is an area of disagreement among keepers and breeders, but it often boils down to a basic cash-influenced decision for the homesteader. In the spring when both grasses and freshenings are abundant there will be milk overflowing the pails. At that time it pays to feed it to the growing kids, rather than making a cash outlay for a second-best type of foodstuff for the young stock.

The labor involved with hand-rearing is extensive but usually fun. There are bottles or pans to be sanitized, prepared, and warmed for each feeding, and then well scrubbed afterwards. The kids drink their milk quickly and that part of the process is downright, capering fun. Do not overburden yourself if you work away from home or have other daily off-the-land commitments, or handicaps such as the lack of running hot water to scrub up bottles and utensils. In such situations, a keeper will get more joy out of kid-rearing and small dairying in general by not hand-rearing all the kids.

Plan to have the separate kid facility readied well before freshening, if you are going to take the kids away. Assemble the necessary feeding equipment. I have successfully used the soft-nippled, disposable, plastic-lined baby bottles as kid nursers, and I have successfully used shallow, pint-sized pans. Success in hand-rearing for me is characterized by ready acceptance on the part of the kids and by the lack of major health problems. Many people feel there is more gulping and digestive upset with pan-feeding than there is with bottle-feeding, but that has not been my experience. When I ran out of bottles I had to resort to feeding some of the kids on pans. Whatever the choice, it may be necessary to teach a kid to relate to the bottle or pan. This is done by gently dipping the kid's mouth and nose in the warmed milk to get it started on the pan, or by dipping the bottle nipple in milk so that the kid gets a taste and tries to suck.

Once you have decided upon the approach appropriate to your own situation, you can give some thought to the question of excess bucks. A mature buck is an extra mouth to feed and often a handful of trouble unless his presence is justified as a herd sire. And very few bucks need to be saved for breeding purposes. It is best to settle your mind on this before there is a barnful of rambunctious, endearing, heavy milk-glugging scamperers on hand. Only a professional breeder needs more than one stud in the herd, and the household-scale keeper probably does not need even one. So plan beforehand *not* to keep your buck kids.

Ideally, only bucks coming from well-documented, high-producing, "milky" families should be kept by anyone. We are currently far from that ideal. We would not have such a surplus of undistinguished, superfluous bucks as we now see if more keepers were thinking ahead. The newspapers hereabouts are full of advertisements placed by people who are trying to give away purebred bucks of all ages and breeds. Plan to sell your extra bucks to a game farm or to the meat market at three or four weeks, or to castrate them and raise them as eventual meat animals or herd companions. Some keepers drown their bucks at birth. All of this sounds very drastic and extreme to new homesteaders or to vegetarians, but after six years of desperate phone calls from new goatkeepers and of observation of surplus bucks, I have evolved a rather hard line on the issue. Facing the implications of the problem early prevents unplanned sexual marauding among the does, makes life easier for the goatkeeper and saves the bucks from later abuse and neglect.

Up to this point we have been talking of kids at twelve weeks and yet have not even seen them through the birth canal. I adopted this order to stress points needing forethought and advance planning. Panicky responses to kidding may result when the physical setup is not prearranged. I have also found that beginning goatkeepers alleviate more anxiety by concentrating on barn arrangements and talking out the *pros* and *cons* of the differing methods of kid-rearing, than by memorizing the physiology of delivery and worrying about possible perinatal problems.

Kidding

Many chapters on kidding tend to stress potential physical problems encountered by the doe. This approach does not correlate with my own observations. Given sun, consistent exercise, and a good diet not overly rich in grain as basic prenatal care, most does, even first fresheners, need no intervention or human assistance in kidding. Lack of daily, full-bodied exercise and direct sun, and an excess of grain seem to account for the large number of kidding problems routinely chronicled in goatkeeping literature. If you feel trepidation regarding a coming freshening, contact your local livestock veterinarian or the oldest, most experienced goat breeder in your vicinity, and make prior arrangements with that person to back you up should some difficulties arise.

You may notice your doe behaving in extraordinary ways on the day of delivery. She may refuse her grain at one feeding. She may seem distracted or a little blurry in her responses. Most frequently, she will pull back from the herd, either declining to go out with the other goats in the morning, or hanging back from the herd as the others graze and move along their usual circuit. Some goats get fretful, but most of mine have become withdrawn and quieter than usual. Without anthropomorphizing too much, I think it can be said that most does seek a situation of quietude and security.

There are several physical signs of kidding you should look for. You may see distension and changes in the vulva prior to kidding, sometimes as early as three to four weeks before the delivery day. Mucous strings or other discharges should be closely monitored and checked by a veterinarian if necessary. The state of the udder gives good alerting signs that indicate how close to time the doe may be. The bag will change from the slack, drooping condition characteristic of the dry period, to a tight, often shiny, filled-out vessel. Milk may be found encrusted on the nipples. Some heavy producers get so full and taut they may need to be milked out prior to delivery. However, colostrum doesn't flow until the doe gives birth. First fresheners' udders develop quickly in the last month of the gestation period and the udder shapes up rapidly the week prior to kidding, becoming full, tight, and glossy. The best way to keep track of these changes is to feed your does their grain on the milkstand twice a day. This practice accustoms them to the stand and gives their keepers a twice-daily checkpoint. By checking often you become alert to the approaching birth, and familiar with general health of the udder. Fluid retention, possible inflammation, lumps, or congestion can be spotted before they turn into major problems. The stand introduces first fresheners to the milking routine and maintains the old-timers' rhythms in the dry period.

Many does seem to carry their kids visibly lower a week or two before the birth. You can often gauge the approach of the specific day by observing or feeling the does' spinal line and rump area. Do this consistently over the last month; setting up a basis for comparison allows the changes of the last weeks to become more evident to the keeper. When the kid(s) move into prebirth position, the does' pelvic bones seem sharply prominent and the *pinbones* decidedly raised.

All of these signs may be present, or only a few, and sometimes none at all. Acquaintances of mine often jokingly tell me about the veterinary college quiz stumper: can you authoritatively say whether this goat is in-kid, and how have you reached your conclusion? And indeed, I have had several mistaken opinions from professionals. Once a doe I was boarding was examined and pronounced not in-kid by the examining veterinarian; six weeks later, she delivered a healthy doe kid late at night, unexpected and unassisted. Often, large-framed does that are carrying one kid may not even look in-kid, so they confound the experts until the last day or two. Hang in optimistically if you are sure of the stud service date and have seen no further recurrences of estrus. Rely on your daily milkstand inspections for perceiving noticeable side bulges, independent internal movements (kicks) and signs of udder development.

As the day draws nearer, breathing patterns change. The doe often handles her stresses by a series of panting breaths. This breathing change is intensified in older does and in warm weather. Older does I have observed spend much of their delivery day lying down, getting up only after the first kid is fully out in the world. They clear the kid's facial orifices of mucus, lick it, then lie down to deliver the next kid(s). First fresheners are often rather dazed by and during their first kidding experiences. Their breathing will become short and quick. Contractions will be visible and the birth moves fast. The kid is often up and about within five minutes of the appearance of the head.

Normal presentation is characterized by the appearance of a red balloon, the water bag, and a head first, front-feet-forward position. Goats will often correct a breeched (reversed) presentation by themselves. Aid the doe in this situation by providing a slope in her stall. Should your doe have to reposition the kid to create a more normal and comfortable presenting position, a high slope will enable her to place her rear legs higher. An artificial slope can be created against a wall by piling up a couple of old feed bags filled with litter or sawdust. Some does may push their heads against this contrivance.

Once the initial signs have appeared, the delivery should be effected within fifteen to forty-five minutes. If your doe is laboring for forty-five minutes and her first kid is not delivered and standing, get experienced help. Most does expel the afterbirth promptly; some does will immediately eat it. This is a protein-supplementing and sanitizing practice, but if it offends the keeper or if the doe makes no move to eat the placenta, remove the afterbirth from the stall. Count the afterbirths that are passed; there should be one for each kid delivered. Does that retain placentas may require veterinary attention.

Postnatal Care

After the kids are delivered, the keeper should clean off the top layer of stall bedding — it will be drenched and slimy — and rebed the stall. If you place bedding containing an afterbirth in your compost pile, dig it in deeply and cover it well with layers of other material. Handled otherwise, it will attract all the carnivorous predators in the neighborhood. Some people I know use sawdust to bed-up kidding stalls or add it as an especially absorbent top bedding layer. Other keepers who use hay instead claim that the acidity of the sawdust creates a more comfortable breeding ground for unwanted types of bacteria.

All you need do during the first few hours after the birth, if you are leaving the kids with their dams, is to see that the newborns get to the dam's teats and down to the business of nursing. This is essential if colostrum is to get into the newborn's system as soon as the kid is ready for it. After giving birth many does will accept an offering of warm water; some will not eat anything till the next regular grain feeding.

When taking the kids away, I dip the umbilical cord in an antibacterial iodine solution, but I let the dam tend to the severing. I do not cut or handle the cord. After separation and dipping, the best treatment for the cord, a remnant of which will hang loose for up to three weeks, is to get the kid out in the sun every day and to avoid touching the cord. It is a good entry point for infection.

As a general good health measure, provide for the kids a means of access to direct sun. I have seen many kids kept indoors in crates and boxes because their owners had no other places for them. Don't deprive your kids of sunshine. Hardly anything equals the sun's antibacterial action, its contribution to growth, or its ability to increase resistance to infection. The sun is available, cheap, and complete. Even a bit of winter sunning is recommended for out-of-season kids if daytime temperatures are not too extreme. Kids separated from the dams are in a generally more artificial set of circumstances and do not have twenty-four-hour mother vigilance, and the dam's assiduous tongue to cleanse them. A converted playpen or a garden enclosure makes a convenient sun-pen. Kidding season comes long before planting in my area, and so I have been able to turn the young kids out into the garden and the dams into the regular pasture, insuring both separation and sun. The kids have no memory of having been in the garden the next year, if they are still around the farm. Once the kids are well established on the bottle or pan and look to human hands for their food needs, they can be run with their dams in the regular pasture.

The keeper should be particularly observant of the excretory patterns of separated kids. The quality of the stool will gradually change from the orange yellow clump characteristic of the young animal to feces that resemble true goat pellets, only smaller. Scours and coccidiosis are problems often encountered with young kids, and the keeper should watch for stool loosenings or other diarrhea-type bowel movements.

Appetites increase rapidly. Feedings must initially be small and frequent. Appetite and capacity for food will vary substantially among kids as they develop, though they will tend in most cases to equalize by the time the kids have reached two weeks of age. By then, on the average, kids will ingest approximately two quarts of milk a day distributed among three feedings. They are also nibbling hay at this point, and I begin to offer them grain. Interest in grain picks up around four weeks, depending upon the growth achieved. I either continue some milk feeding till twelve weeks and introduce a half pound of grain to the diet from the sixth week on, or go completely over to grain between six and eight weeks. The amount of grain fed will vary with the quality of hay being provided and the availability of fresh grazing. When there is substantial fresh grazing, the kids come off milk early and get fed only small amounts of grain. If the kid is being weaned and reared through the winter, it may need one-half to one pound of grain per day, plus its free choice of hay.

Dehorning

Horns must also be given advance consideration. The easiest, least traumatic, most effective, and most sanitary dehorning procedure is to use an electric dehorning iron on the kid's horn buds before it is one week old. Four or five days seems to be the preferred age for dehorning, though there are people who disbud on the second or third day after birth. Early disbudding may be preferable for buck kids going on to a breeding career. In order to do a really complete disbudding on a buck kid, do it early. The hornbud area is larger on the buck kid and develops earlier than on most does. Since buck horns grow vigorously, a disbudded buck sometimes gets a second growth of distorted or modified horn tissue called a "scur."

Dehorning is often glossed over in books because it is best to learn the whole procedure by way of personal demonstration. Have an experienced person coach you the first few times! Such a person is not always within reach so if you can't find one, I hope the following description helps.

You'll need help! Enlist a friend who does not flinch at the outcry of a sweet young kid or at the smell of roasting meat. Have this brave soul hold the kid while you work on it, placing the kid's four legs in the well of his/her lap, and keeping hands free. One hand can hold the kid across the face; a hand across the mouth and nose produces a firm hold. The other is free to restrain the trunk or body. Avoid holding the kid around the neck or throat.

The person doing the dehorning should clean the copper tip of the electric dehorning tool well with alcohol, plug it in, and set it aside to heat up. Then with pet grooming clippers shave the top of the kid's head around the entire horn-bud area. My clippers need a drop of oil on the blades each time they are used to get a good close shave. The

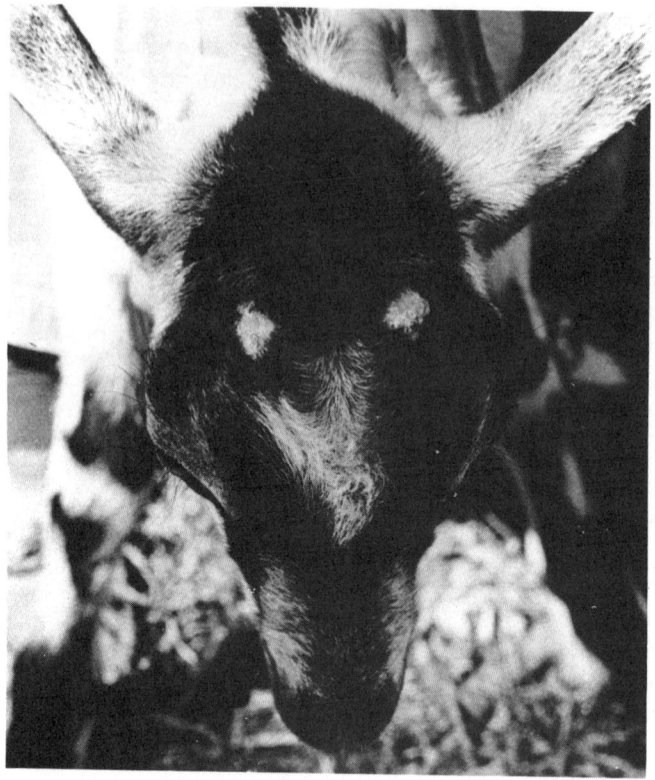

A recently disbudded doe kid. The round white spots indicate a clean, well-healed disbudding job.

sound of the electric clippers may startle the kid, so turn them on for a few seconds prior to approaching. Then come up close and shave the hair as closely as possible.

Wipe the shaven horn-bud area with a cotton swab dipped in alcohol. Apply the hot iron firmly to each horn bud for ten to twelve seconds. The kid will flinch and squirm on contact, but briefly. The iron is actually cauterizing the area, and most of the disbudding is achieved by this initial contact. Work over any protruding, ridgy areas left in or around the horn bud. When you are done, the bud areas should appear as two bright coppery-colored circles with a lighter, whitish center in each. The entire procedure takes twenty to thirty seconds per horn bud, counting the initial ten-to-twelve-second first application and any touch-up you may have to do. I apply no creams, ointments, or dressings to the cauterized area, but put the kid out in the open air and sun as much as possible afterwards as prevention against infection. Cleanliness of the implements and careful, thorough alcohol swabbing help to prevent problems.

If you pasture your goats with horses or keep them on acreage that has had horses on it, inoculate all your animals against tetanus. Tetanus bacteria proliferate in manure and find horseshit particularly homey; these bacteria can be dormant for a long time. I have heard unsubstantiated tales of kids contracting tetanus after being dehorned, but have not had any direct experience of it. However, the type of wound that is created by disbudding with an electric iron is a cauterized surface burn, not the deep puncture wound that provides tetanus bacteria with their favored environment. *Clostridium tetani* grows in the absence of oxygen.

Another method of dehorning involves putting rubber bands, or elastrators, around immature horns. Avoid dehorning young stock with elastrators unless you innoculate against tetanus several weeks before applying the elastrators. Personally, I feel the rubber banding method keeps the kids in great discomfort for a lengthy time period and runs much greater risks of infection than electric disbudding.

When looking for help with dehorning, go to your nearest dairy goat club and contact its members. I visited many herds in my region and went directly to the owner whose animals had the neatest dehorned heads I could find. He was delighted to start me off right. It may turn out to be more difficult to find someone to help you restrain the kids. People with a 4-H or a farm background are your best bet because they have often had prior experience with other animals being dehorned. Your local veterinarian may be a resource person for dehorning kids, but he may not. The veterinarians in my area have had more experience surgically dehorning mature goats than they have had dehorning kids. Veterinary charges may be prohibitive for a home visit, but kids are easily taken to an office. In the long run it is better to teach yourself the simple practice of early dehorning, for a keeper is less likely to delay dehorning if he does not need to fit into a veterinarian's busy schedule.

The reasons for dehorning are many. Most important, it averts damage within the herd and the possibility of accidents occurring to humans. Goats do not hesitate to use their horns on each other. They are very territorial about their food, their grazing prerogatives, their social status, and their kids. An animal with a fully grown set of horns is beautiful, imposing and formidable. Once you keep more than two or three goats, dehorning becomes imperative. They can injure each

An elderly goat of strong character, unknown but mixed origins and enormous productivity.

other's udders in their everyday territorial combats, and can get mean tempered and bullying toward kids other than their own. A dehorned kid is also more easily sold, since people are concerned about their children's safety and that of visitors unfamiliar with goats.

Initially, I kept fully horned goats. I had a herd of mature, horned goats and a set of big-breed-fancying dog lovers as immediate neighbors. Horned goats can defend themselves against large dogs to some degree, but dehorned goats are totally defenseless. Now that I have a large pasture defined by a three-wire electric fence I have begun to dehorn kids in their first week. The fence deters most dogs.

Routine Postnatal Concerns

If you plan to record or register your goats, they must be tattooed with appropriate identifying marks, the latter signifying year of birth, herd initials if you have them, and an individual identification number. Tattooing does not have to be done as early as dehorning. There is no physiological imperative surrounding it. When breeders have many kids to record or register, they tend to tattoo early. The technique is simple and there are instructions in every tattooing outfit.

Worms are a critical health problem, particularly in the newly fresh doe and the young kid. Severe infestations can deprive the dam of vitamin B_{12}, making her prey to a large number of other problems. Worms can successfully compete for needed nutrients in the system of the growing kid, robbing it of a sound start and good growth rate. Since the majority of freshenings in my area occur in the spring, in warming, moist weather that plays good host to worms and their larvae, I often worm the herd at that time.

There is all kinds of advice on the signs to watch for to spot worminess: erratic or voracious appetite, clumpy or mucusy bowel movements, and my favorite, "general unthriftiness." Well, it takes an experienced eye to note such signs or to decide what is unthrifty. Do not wait for signs to pile up. In the spring scoop fresh pellets from each goat's stall into a plastic bag and take this labeled fecal sampling to a veterinarian. In this way every goat in the herd can be checked for parasites. This can be done by the keeper if he or she has veterinary guidance, a good microscope, some experience, and appropriate texts. The microscopic check will reveal which parasites are in evidence and whether the count is high enough to warrant worming. Then each doe can be dosed according to veterinary counsel.

Preventive measures should always be employed. Chief among these is using a large enough pasture or rotating pastures so that the goats do not keep reinfesting themselves with larvae. If you do not have two pastures fenced in, tether the goats for two-week grazing stints in new areas. Use garlic, rue, wormwood, and other bitters as vermifuges in the ordinary diet of the herd. After much trial I have found garlic useful as a preventive. Many ruminants gorge themselves on wild onions and wild garlic in the early spring; this is probably a self-healing measure. Although garlic keeps infestations down, I have not found it sufficient to rid an animal of heavy worm infestation. By using rue, wormwood and garlic as additives to the does' usual regimen I have been able to keep the chemical worming down to once or twice a year. Feeding two or three cloves of garlic a day for four or five days will not produce garlicky milk if the doe gets the garlic at a feeding twelve hours prior to the next milking. Some keepers are reporting success from using small amounts of diatomaceous earth in the stall bedding to prevent parasite problems. These minute inorganic particles desiccate larvae. I have not tried this out of reluctance to handle the material, which bears a warning against ingestion on its label.

Many keepers find they have to worm in-kid does when winter confinement or other factors produce a high parasite count in their fecal samples. Currently, the wormer that is regarded as safe for the in-kid doe is thiabendazole. Many breeders routinely worm all their goats with this substance every six to ten weeks. This has always seemed extreme to me, since so many other microorganisms get cleaned out in the general purge. The goat probably needs a reasonable proportion of these denizens or they wouldn't be there to begin with. There now seems to be a move afoot to moderate the extreme chemical worming programs by educating keepers to the cycles of parasite reproduction and the tolerability levels of the host animals. Today, one even meets with statements that the wormless goat is probably not even efficient or economically desirable. The household keeper usually has no need to resort to year-round chemical worming since she or he keeps fewer animals close to each other and can more often allow access to outside grazing. It is good to keep in mind that the more confined animals are, the more chance of worm infestation, and the greater the need for more frequent chemical treatment.

Once the doe is in-kid, it may become difficult to keep her hooves trimmed. She may resent trimming and put up a stiff kicking bout. Do not risk an abortion by trying to force the issue. Try to keep hooves trimmed back in the first two months of the gestation period. If the doe resists and kicks, let the trimming go until after freshening. Uncooperative behavior can usually be avoided by accustoming all the goats to a foot trim, or at least an inspection, once a month from the time of birth. Untrimmed, Turkish-slipperlike hooves can harm the doe's legs, udder musculature, and spine. Start kids off right by trimming and checking their hooves regularly. Even keepers who have rocky pastures which pare hooves down naturally benefit from routine inspection of their stock's hoof conditions. If the pastures are not rocky or the goats are stalled, yarded, or indoors for a long stretch in winter, line the herd up for regular pedicures.

After freshening, the doe will pass some bloody matter for up to two weeks. Clip her neatly around the tail and udder area. Brush her daily. If the postfreshening weeks coincide with the onset of warm weather there may be a substantial hair loss.

If hair loss is great, check immediately for lice, particularly if it has been a long, confining winter at close quarters. Goat lice are hard to see. One way of spotting them is to place a hand on the animal's underbelly or other suspect area and keep it there for five or six minutes. The heat generated will encourage lice to surface and move out of their hiding places. Then look for tiny, grayish, almost transparent critters. They have favorite places to hang out — around the moist orifices of the body or up in the pockets of hair and muscle around the

(continued on page 72)

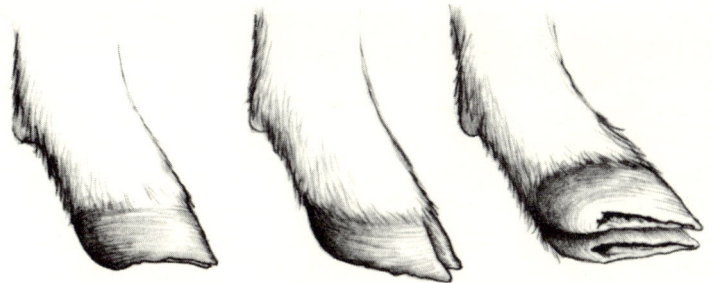

Hooves Before Trimming

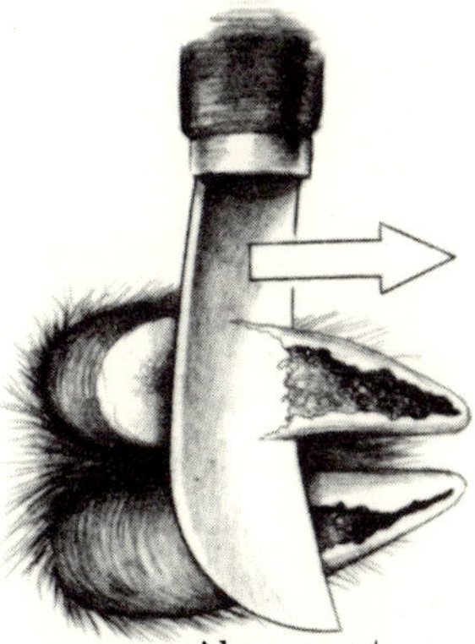

Always cut
from heel to toe.

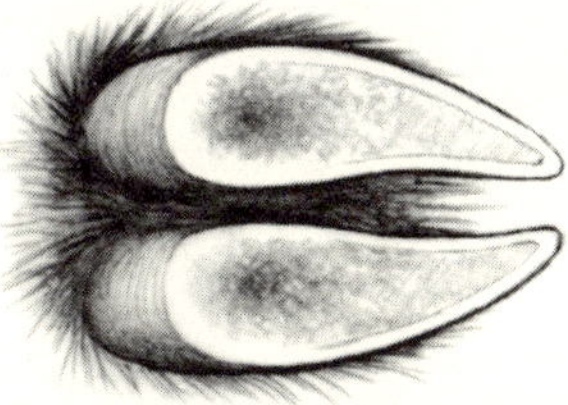

Heel   Toe

Bottom View
After Trimming

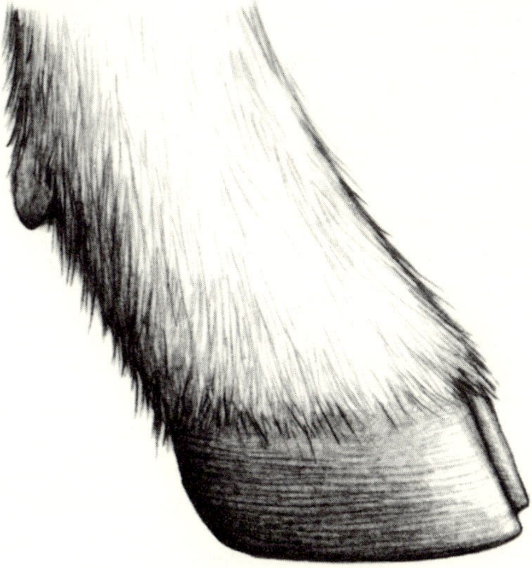

Side View of Properly Trimmed Hooves

Young buck with "Turkish slippers"—overgrown hooves.

Overgrown hoof being trimmed with horse nippers. The arrow indicates the overgrowth.

*Two persons are often needed for gentle restraint during trimming.
Note the first aid preparation at hand in case the blade slips.*

Trimmed, squared-off hoof.

udder. I have not had them lurking around my animals long enough to be able to give more data than this. Exposure to sunlight and the outside vicissitudes of weather seem to take care of them when only a minor infestation occurs. Lavender oil, rotenone (called derris powder in Great Britain) and pyrethrin sprays have been successfully used by acquaintances of mine to combat lice infestations. They should be used with veterinary supervision. Milk from lactating animals that have been treated with rotenone or pyrethrins should be withheld and discarded. Lavender oil smeared in key places is effective and has no toxic side effects. Juliette de Baïracli-Levy recommends the scattering of quassia chips in the stalls as a preventive measure against lice.

Patches of flakiness, dry skin, and some loss of hair may be encountered in does during postparturition when they need more fat-soluble vitamins. When vegetable oils skyrocketed in price several years back, feed companies as an economy move reduced the amounts of soy and other oils mixed into commercial dairy rations. After a very long winter and the stresses of a pregnancy, a doe may show some of the signs of fat-soluble-vitamin deficiencies. I have dealt with this by routinely adding one teaspoon of wheat germ oil to my does' grains once a day. After two or three weeks there is a complete reduction of hair loss and little or no skin dryness. Once spring arrives and green grazing is again available, oil supplements become unnecessary because there is sufficient vitamin A and E in the fresh greenery.

A combination of worm infestation, pregnancy, and lack of outside access can result in serious vitamin and trace mineral deficiency, particularly in cobalamin, vitamin B_{12}. Newly fresh does suffering this deficiency will have acrid, unpalatable, off-flavored milk. Worm the doe, and get her into the sun to graze, even if fresh grass or bark and conifer needles are the only available grazing. Dissolve a mineralized salt block in two gallons of water. Further dilute the salt-mineral solution and add it to the doe's drinking water, a small amount at a time, over the next two weeks. Always taste your doe's milk and check its quality visually by inspecting the strip cup whenever you milk. The milk functions as an index to the condition of the lactating doe, both when she is newly fresh and further on into the lactation. Sweet, fresh-tasting milk of a fine white color that has no textural oddities is the sign of proper milk goat health.

CHAPTER 7

FEEDING

An older man of my acquaintance says that the basic rule of dairying is that you get out of her just what you put into her. I have a firm belief in the efficacy of putting in affection and care. It helps, and may even be the determining factor. I have, however, met many homesteaders by now who are putting in much love but getting nothing in the milk pail. A group of folks I knew were supporting eleven mature does, but receiving a total milk yield of eight pounds, or four quarts, per day at the seasonal peak. They loved their animals dearly, but eventually sold off the goats because they needed milk. To avoid such paradoxical quandries it is necessary to pay attention to the other input factor, the one my older friend had in mind, the goats' food.

The standard American text on feeding animals, Morrison's *Feeds and Feeding,* does not discuss goats.[9] David Mackenzie is excellent but British-oriented, and originally published in 1957. Current pamphlet and periodical literature is voluminous but it still is not specific enough, and it bespeaks a generally nonempirical, derivative approach. The goat is often regarded as a little cow.

The Goat on Pasture

The essence of my empirical and direct approach, by contrast, is to watch a goat graze. She adores variety. It is an endless cruise; a nibble here, a tuft there, a strip of this and a peel of that, a gobbling up of the next thing. Some days it is a mad dash to the burdock patch, a voracious tearing of leaves, a rapid ingestion like the suction of a vacuum cleaner. The next day the perimeters of the burdock patch are sniffed and snorted at and definitely scorned. Repetitions of this seemingly fickle pattern are endless, but the repetitions are the rule. Feeding rule number one for me is to provide the goats with as broad a selection as they need in their environment.

The goat's fickleness is partially explained by her physiology. Her mouth is tough and adapted to taking in the crudest, thorniest, most abrasive, and bitter-tasting collection of wild vegetation that grows. She also has an enormous capacity for roughage intake given her size. But she'll only eat the choicest portions of the wild vegetation. You can see her lopping off flower heads and seed tops in her meanderings. Feeding high, eating the tips and new growth of plants or the first few leaves of just about everything, helps protect her from ingesting harmful worm larvae that lurk lower down on the ground. She passes on to the next tidbit because there are only so many tops to lop off and eat. Her pattern makes her into some sort of four-legged lepidoptera, making just a brief visit to one plant and then a passage on to the next, and the next, and the next.

Cows may graze on clover, but to a goat it is not especially enticing. Our goats also ignore ryegrass until late November when there are few other fresh greens. And often in the late fall when the choice is between perennial ryegrass and fresh pine needles, the needles get sucked up with great relish and the rye is left to languish further. Plants that I have seen the goats consume in quantity and with seeming pleasure include:

blackberry brambles	Chinese elm
wild roses	weeping willow
sumac	maples
day lilies	honeysuckles
wild onion	spirea
burdock	forsythia
plantain	wild carrots
dandelion	mustards
chicory	asters
sorrel	all types of sunflowers

wood sorrel	vetches
curly dock	clovers
thistles	ryegrasses

Last in the goat's preference are clovers, ryegrasses and vetches. In the winter the goats eat pine, cedars, fir, some spruces, and probably many other evergreens that I just don't happen to have on my acreage. Fruit trees are also a great favorite; guard them well. Despite the appearance of chokecherry and milkweed on numerous lists of plants toxic to goats, I have seen both consumed by goats with no ill effects. Perhaps many of the listed toxic plants are detrimental only at certain stages and times of the year.[10] Spreading myrtle, periwinkle or vinca minor, is definitely not a good plant for goats and ours scrupulously avoid it at all seasons. Mountain laurel, azalea, and rhododendron are also dangerous. Experienced older does that have had a lifetime of outside foraging experience carefully avoid these plants. A young doe out on her own for the first time with an even younger kid for company may stray and eat these toxic plants. So if your shrubbery includes dangerous plants keep your goats far away from these plantings. Uprooting dozens of various plants from a pasture may not be feasible, and you can't count on getting every one of them. The best precaution is not to let the stock run free without an older, experienced doe leading the foraging parties.

Goat Fodder

Watching goats' peregrinations and feasts in the wild, clues you in to what to grow, cut and haul for them. Jerusalem artichoke and Russian sunflower stalks and foliage, all kinds of brassicas from October on, and fresh comfrey are among our goats' favorite cultivated crops. We take certain parts from these for our household table use, such as cabbage and broccoli heads, artichoke tubers, and the comfrey root, but the remainder gets cut and carried to the goats for bad-weather supplementation. They are glad to get the outer leaves of the brassicas. Jerusalem artichokes are enormously productive of goat fodder, a by-product we had not anticipated when we planted them for our own use. And we learned to eat comfrey leaves by watching our animals enjoy them, having originally used only the root for ourselves. Now we cut comfrey leaves fresh for salads and salad dressing, use it both fresh and dried for tea and, of course, feed it to the goats.

Since our goats forage and consume a large amount and a great variety of deep-rooted plants, we have not run into major mineral deficiency problems. We supplement forage with mineral-supplemented grain feed and a large mineralized salt block in the pasture and in the stalls. Dependence on the stall blocks should not be complete because some goats ignore them and others knock them about. Friends of mine dissolve the mineralized blocks in water, dilute the mix, and then add it to the goats' regular drinking water. I have had only one goat that relished drinking the slightly salted water. I now prefer to rely on the pasture blocks because I see that the goats visit them regularly.

Hay

If you stall and yard your goats, your dependence on good quality hay and on planting, cultivating and hauling fresh greenstuffs will assume great importance. This has been done for centuries in European towns and villages where backyard dairying is quite common.

Hay is important to all keepers. Having started as a total hay ignoramus, I'm in no position to present a treatise about feed hay. But because of necessity, I have gone through some rapid learning experience, and I have some information to pass on.

During our first two winters, we fed the goats what is euphemistically called around here "hay from native American grasses." Our goats ate it, survived, and even did well. We averaged sixty-five forty-pound bales among the four goats over the six-month period of little or no nutritious outside grazing. Of course, the goats went through three or four bales a week when the nighttime temperatures sank below zero and the days permitted no outside bark foraging at all. During other milder weeks, only one or two bales were consumed. This is all very general because the age, size, and condition of your stock are variables to consider in feeding, in addition to the weather. The quality of the hay we secured the first winter ran in a range from worthless, browned-out, zero-nutrient grass hay to presentably green, decently cured but late-cut mixed grass and clover hay. After the first winter we realized that the few surviving local small general farmers had to keep their better hay to feed their own dairy animals. If they sold any hay at all, it was not very good hay, and the prices were steep because they still had to cover labor on the haying. If you announced that the hay had to be feed-quality and that you were feeding it to milk goats, that information was characteristically met by gentle laughter and offerings of weird hay that could not be fed to any productive milk animal. Even the most well-intentioned local farmers definitely had the notion that a goat would and could eat anything. To compound the problem, we had only book learning and could not recognize the major legumes or grasses or ask the right questions.[11]

What we did was to bring sample bales home for the goats to review. If Mackenzie scorned, discarded, or trampled it, we accepted her decision. What she preferred were the bales that included green-tipped, comparatively coarse-stalked plants. These could be clover, alfalfa, heavy brambles, chicory, or pigweed, plus any grassy things that had seedheads. The goats ate all these things with equal relish, as long as it was green, dry, and not dusty. Whether they derived any nutritional value at all from some of that hay is questionable. We fed them a 14 percent protein-content grain ration at the rate of from one to two pounds a day, depending on age and whether or not they were in-milk or in-kid. We knew that the hay was yielding fodder for generating warmth but probably little maintenance protein. It was a mild winter for these parts, but we went through lots of hay.

The next year we decided we either had to produce our own hay or buy the winter's supply from a hay-producing farm. The decision was to buy, because we lacked equipment and an adequately improved soil at our place. We trucked twenty-five miles each way to a professional hay farmer who always had hay to sell because he grew it as his primary income crop. We purchased too late to secure an early cutting, but we did buy a well-cured, good green-content, native grass hay. It included some legumes and was heavy on the strange assortment of pasture weed plants our goats liked to eat. And this farmer explained to us the mysteries of hay. The best feed hay is that made from a well-cured first cutting of a leguminous planting. It provides productive animals with the highest amount of total digestible nutrients. This type of hay, when it includes a substantial legume content, provides protein as well as roughage.

It is a moot point whether the dairy goat, even a stalled doe, should be fed a protein-rich alfalfa or clover hay exclusively. The debate can often be resolved for you by the vicissitudes of the haying season and your own financial predicament. Prime, well-cured legume hay from a first cutting will often be very expensive and in great demand. Usually a mixed grass-and-legume hay is much more available and much less expensive. I follow the orientation of David Mackenzie who strongly inclines toward grass mixtures and away from hays with high legume content. Moreover I have my herd queen's strongly demonstrated preference for coarse, weedy mixes and I can graze the goats out most of the year. I don't expect phenomenal milk production through the depths of winter and I never seem to have the money for the fanciest of hay. Therefore I have not purchased or fed such hay. If you keep stalled goats you may opt for alfalfa hay because it offsets the need for large amounts of grain protein, but it may be hard to locate a constant supply in your area.

The other factor we learned about while reviewing the various legumes and grasses, and seeing what they looked like and which cuttings of them yielded the highest nutritional value, was the variations in curing method. I had assumed the best curing was sun curing in the field. Our hay supplier cuts the planting and leaves the cut hay in the field only a short time. He uses fossil fuel-powered blowers to complete the curing. Curing entirely by sun in the field reduces the nutrients available in the final bale of feed hay. An ecological dilemma is presented the small goatkeeper, given the fuel energy needed to cure the hay with blowers. You either trade some nutritional content and buy or produce sun- and field-cured hay, or you pay the financial and ecological tariff for hay cured with dryers. If the higher nutrients in hay enable you to feed less grain and still achieve the milk production you desire when keeping confined goats, the ecological tariff is probably balanced by the decreased grain consumption.

The third winter we compromised in all directions. We purchased a higher-legume-content hay, but from a field-cured late cutting. I also made an attempt to barter pottery for hay, via *The Mother Earth News,* but it produced no tangible exchange. The goats went through this higher-legume-content hay at a slightly slower rate than they moved through the grass hays. However, I have no decided conclusions about reduced grain consumption. I suspect that the field curing and late cutting produced a nutritional content that was not too different from the dryer-cured grass hay of the previous year. My only solid conclusion in that goats will waste alfalfa-bromegrass hay with the same nonchalance with which they waste bramble-pigweed-clover bales. Even with the cleverest of hay feeders built especially for their wasteful ways, goats throw a lot of hay around. Once it hits the floor they scorn it.

The next step is for us to produce our own hay. It involves manuring and several years advance preparation of our scruffiest acreage, an old parking lot and access road area, and making arrangements for renting or borrowing equipment. The latter is hard to do around here because of the very small number of still-active farmers and the need everyone has for the same equipment at the same time. Folks I know in areas that have more ongoing farms have been producing their own hay from their own small holdings by mutual exchanges and arrangements with their better-equipped, slightly larger-scale farming neighbors. It is not the fanciest or most nutritious hay, but it is unchemicalized, produced by home labor input, and free. Home-growing also reduces hauling time, labor, vehicle wear and tear, and fuel consumption.

Grain

Grain costs represent the largest cash outlay in the goat enterprise. We buy ours. It is not organically raised, and that is our strongest objection to buying grain. The best we have been able to achieve is the purchasing of a grain ration that has no additives, preservatives, urea, or antibiotics. Supposedly, the USDA monitors dairy rations for pesticide residues. We don't count heavily on this — the levels of tolerance the USDA accepts are not mine. Until we are able to obtain organically raised grain our milk will not be perfect. So far, we have only achieved the production of a milk that is discernibly fresher, with a higher mineral and vitamin content than what we were drinking before. It is certainly less processed, less loaded with substances I regard as foreign to dairy protein: antibiotics, emulsifiers, preservatives, and reconstituted solids from older processed milk. We would like to evolve toward producing milk that is also free of pesticide residues, but for the interim since we have no way of producing our own feed grains, we must be satisfied with the gains we have made over our prior sources of dairy protein. If you do raise your own grains, a 50 percent oat, 50 percent corn ration; supplemented with molasses and minerals is usually adequate.

We feed a 14 percent protein content mixed-grain ration. Young stock gets half a pound a day distributed in two feedings when they are on full grazing, and up to a pound a day as they approach the yearling stage and pass through the winter. The does in-kid or in-milk get up to two pounds, depending upon time of year, their condition, and overall size, possibilities for outside grazing, and where they are in their reproductive cycle. We vary the amounts constantly on the basis of season, the type of hay being fed, and the metabolic curve and productive capacity of the herd members.

I have experimented with many, many grain rations. They seem to vary even more than the demands of the goats. Don't settle into a routine and count on using uniform rations. There are seasonal variations in feed content and physical differences in the compounding. The milling is not always consistent. Sometimes the grain is milled much finer than at other times. Sometimes the grain content or mix is changed radically. Once the ration I had been feeding turned up in pellets. The mill had just started pelletizing all its dairy rations. Our goats thought it vile stuff and unequivocally rejected it. Neighbors of

79

ours had reverse luck; their goats seemed to love it. We had to switch suppliers and mix the pellets in with the next batch of old-style grain we purchased. The goats were notably successful at picking out the detested pellets from our new mix and often didn't eat them.

When feeding some very poor hay, we fed an 18 percent dairy ration. You must be careful when purchasing very high protein-content dairy rations; check the label for any urea content. Goats do not do well on urea, and it may even harm them. The labels are sometimes opaque on this score, but the clue is usually some vague reference to "protein from nonvegetative sources." Once you spot that type of phrase, examine the label closely. You will find urea listed in the long line of ingredients, somewhere down in the fine print.

Other points I have noticed about grain feeding center around goat habits and personality. All you have read about goats being creatures of habit and resistant to change is probably true, but don't count heavily on it. Each goat is an individual and has evolved habits peculiar only to herself. In general I would say they do not like dusty, small-particled grain rations. They love corn kernels and will break into a chicken feeder with no hesitation. I try to feed as little corn as possible, since feed corn is one of the most chemicalized grains raised by agribusiness. The bulk of the ration fed is comprised of wheat, bran, oats, small amounts of corn, and soy derivatives. Follow your animals' obvious preferences but give your own convictions some credit too. We avoid mixes with heavy corn content, pelletized mixes, and overly dry mixes, and we even force the goats to some regimes. One of these is the substitution of bran for part of the regular ration during the month prior to freshening. Starting with a small amount of bran, we work up to a half-bran-half-regular-ration blend in the two weeks just before the kidding. This is not very much to their liking because bran is light, dry, and flaky, but my reasons for feeding it are explained in the chapter on kidding. The goats have not yet bought our explanations, but do reluctantly eat the bran when the time comes. I wet down their grain with two tablespoons of cider vinegar during the latter part of the pregnancy so the bran is just palatable enough to be grudgingly accepted.

We also vary the amount of grain we feed when we wish to vary the milk yields. We have actually had to reduce the yield at times because we had no way of using it all. That situation has changed now that we have evolved satisfactory methods of making hard cheeses. When we had too much milk flowing in, we reduced the amount of grain being fed. It was the season of the finest grazing, so the goats did quite well on less. Slowly increasing the grain toward the normal feed level, slowly increased the yield again.

The other period of the year when we used to cut back the amount of grain being fed was during drying-off. If the weather was unusually mild and the milk yield had not dropped off sharply, I cut back about a pound a day on the grain when I started to dry-off the animals. This was a very temporary cutback and aimed at somewhat reducing output while I started decreasing the frequency of milkings. I stopped doing this because I believe feeding less grain at this point can stunt fetal growth.

Knowledgeable older goats, though they love their ration, rarely glut on grain. If you observe your goats and they seem to be attacking the grain voraciously, then erratically, with attempts at breaking into the chicken feeders or the grain storage bins some days, perhaps you should suspect worminess. I do not trust young stock and goats that have just come to us from other places and other lives to be reliable about eating just the grain they really need. These goats may be wormy, confused, undernourished, and crazed for grain. The young ones have very little experience in learning about proper amounts for their systems. As long as an older doe is around making the decisions for the group and disciplining the young ones, they follow that guidance. Little ones on their own can do such things as get into the grain and overeat to the point of bloat danger. And do not depend solely on your stall latches for security on this point. Determined goats are notoriously clever and persistent about breaking and entering. Tie down your grain cans so you can control the feeding regimen.

Our mature does have revealed sudden appetite leaps between the tenth and twelfth week of being in-kid. Last winter one of them battered down her stall door, went to the grain cans, and helped herself to what she felt was an adequate increase. After that we followed the guidance of the does' appetites at that stage and allowed them a slight increase, trying to keep it small because we know that overdosing them with grain concentrates during pregnancy can make for delivery difficulties at kidding.

There is a tendency among beginning goatkeepers to make grain-feeding decisions on the basis of their anxieties about the young ones getting "enough." Once the freshening takes place we relate our grain feeding to the quality of the available grazing, to the number of kids being supported, and to overall considerations of milk yields and the size of the doe. In my area, April dams have grazing available to them in most springs; March dams encounter early, restricted grazing situations.

An old goatkeeper I know feeds one pound of grain for every quart of milk he receives from the doe. I feed much less than that — usually about a pound per quart and a half yielded. Our does thrive and give substantial yields on this amount of grain plus available pasturage. Our older doe averages between eight and nine pounds for three months after freshening on all-day grazing and three pounds of grain, with diminishing yields as the grazing diminishes in quantity and quality. Her peak seems to have occurred the summer I bought her, when she had freshened in a lush June at three and a half years of age; she milked into October at a rate of twelve pounds daily and then on for a total of twenty-two successive months. Our first fresheners average between six and seven pounds for their warm weather period on all-day grazing and two pounds of grain a day.

Our young purebred buck receives one pound of grain over the day plus all the grazing or hay he desires. We give him a slight bit more grain in breeding season on the advice of a twenty-eight year goatkeeping veteran who has kept many bucks and feels that they can utilize a little extra grain bonus when their working season arrives, but that they should not be allowed to run to fat.

A few last words on the grain: we do not weigh out every portion for every animal every time we feed. We do, however, use the same measuring cup always so that the volume is the same, and we check each new grain bag for the bulk-to-poundage ratio. We have experienced so many differences in the rations by now that we have seen that a drier-milled, finer, fluffier ration may take up a whole cup several times before that pound is reached, whereas, one cup of a dense or pelletized ration may equal a whole pound. It is the old story of a pound of feathers taking up more room than a pound of gold. If it has been consistently milled and mixed, each pound of the ration should contain the advertised percentages of protein that the label leads you to expect.

Consider, too, one final aspect of getting out of her what you put into her. Be sure to taste the milk produced every time you milk. Goats eat an enormous variety of plants. When they are out grazing and selecting freely, they may ingest goodly quantities of plants that they adore but that you find you cannot abide when you taste them in your (their) milk. Wild onion and garlic are spring favorites of our goats, perhaps for their medicinal functions as vermifuges. Rather than police them all day or dig out the offending plants, I consign such heavily flavored milk to cooking or baking. It is true that you can sometimes taste the grazings of a milk doe in her fresh milk, but these episodes are brief and piquant.

CHAPTER 8

SHELTER AND PASTURE

The milk goat needs shelter. In the wild, she is clever about securing natural sites that offer protection from the wind, wet, and cold. She is also free to eat naturally, selecting what is appropriate for her age, condition, climate, and general needs, as long as there is herbage to select from. Once she is removed from her wild habitat, housed in a barn, and given a specific pasture or yard, she can no longer exercise natural diet selection, nor has she complete control over any of her grazing circuit. The keeper has to step into the vacuum and create a situation that offers protection and easy management, while taking care of the milk goat's essential nature and instinctual patterns.

The more removed a goat is from her wild habitat and freedom of selection, the more the keeper will have to supply. This can become expensive in labor, feed and fodder hauled, general monetary expenditure, and net energy costs. An example at the farthest end of the scale from the goat in a feral environment is the goat kept in an insulated and heated barn with limited or no exercise or pasture access. In such an environment the keeper is responsible for hauling and cropping all foodstuffs, for the maintenance of an artificially created and sustained "climate" in the shelter, and for any problems that may arise from the goats' not being hardy and acclimatized to the outside world. This is a very costly management scheme. In northern climates the rise in fuel and insulation prices almost precludes this approach, though it may still be feasible for established breeders and persons planning to support their goat enterprise with a sizeable outside income. Persons with less outside income and fewer ambitions about developing a show herd can do quite well with simple shelters that are unheated and uninsulated, as long as they are draft free, tight and dry. A goat sheltered in such housing will be able to sustain winter wanderings and grazings in tough climates.

Lean-Tos, Sheds and Barns

In mild climates without prolonged, deep winters, very simple shelters suffice. Lean-tos and sheds that offer wind protection, shade for the days of strong sun, and a minimum of dry comfort in rainy weather will serve. Shade in times of hot sun is important to the dairy goat out on pasture. That heating system she carries is constantly generating warmth — it doesn't click off in hot weather. She has a high body temperature, between 102° and 103°F. (38.9° and 39.4°C.) normally. Too much direct sun can be burdensome to her. Shade trees in the pasture, lean-tos, or direct access to her regular overnight sheltered is needed.

In northerly temperate climates, you can choose one of two management schemes. In one method, you can plan for eight months a year of outside grazing with short outside grazing spells during the

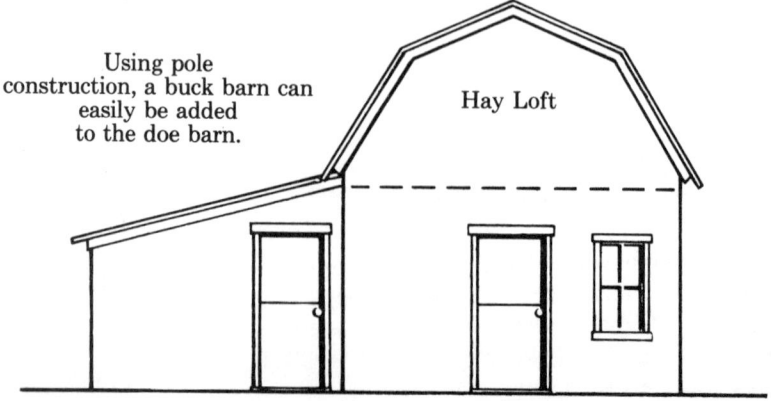

Using pole construction, a buck barn can easily be added to the doe barn.

Hay Loft

deep winter months. Housing must be dry and windproof with windows and doors on the south side of the structure, if possible. Good ventilation and sunlight are necessary. If the goats cannot get outside for a few days, a good southern exposure is helpful. If the door is on the south side, fewer wintry drafts are let in when you come to do the chores. The goal is to provide a shelter that offers the basic physical protections and psychic security in severe weather, and still remains cold enough in winter not to differ too radically from the temperatures prevailing outside the barn. In this way, the does develop hardiness.

My situation provided a good, tight, stucco, dry structure, but the door was on the north wall and there was a window on every wall. We

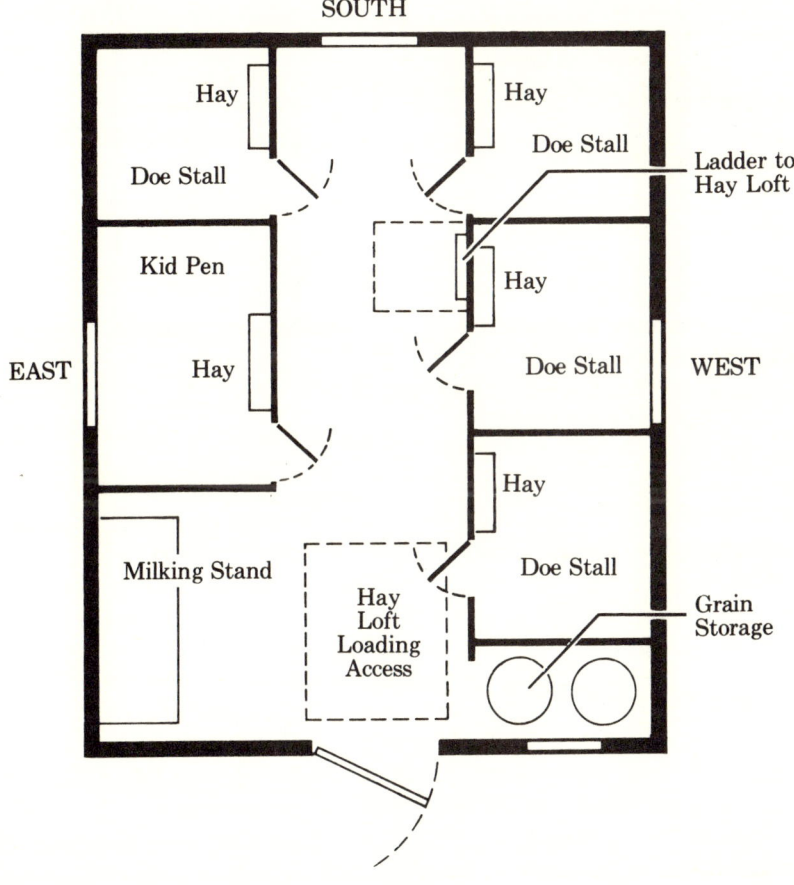

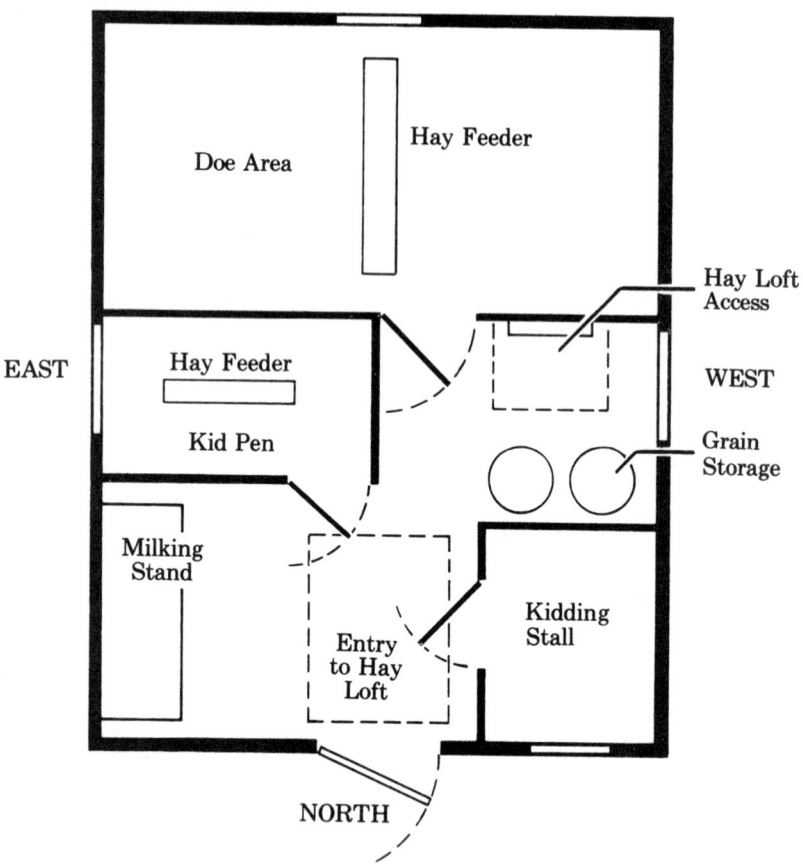

SOUTH

Doe Area

Hay Feeder

Hay Loft Access

EAST

Hay Feeder

Kid Pen

WEST

Grain Storage

Milking Stand

Entry to Hay Loft

Kidding Stall

NORTH

had to take a few compensatory steps because of all the northern openness. We stapled plastic over the windows, except those on the south side, during the winter. We built a storage loft overhead to accommodate hay and to lessen the ceiling height which was fourteen feet at the peak of the roof. Lowering the ceiling helped keep the warmth generated by the animals down where they needed it. When the hay loft (capacity, sixty-five bales) is filled, the hay functions as insulation. The forty-to-fifty-pound bales are hoisted up into the loft by two persons with the aid of an inexpensive one-ton hoist. If necessary, this operation can be performed by one person.

To further insure comfort on sub-zero nights, we keep the goats on deep litter. Left to their own devices and notions of sanitation, goats will bed down on a comfortable bed of their own dung. It is warm, familiar, and easily available. Microorganisms are active underneath this thick bedding, just as they are in a compost pile. The deepest layers of the litter are in a process of ferment and bacterial break-down, generating warmth. We don't muck out from November until February unless there's an unusually mild spring thaw. With her in-built portable hot water bottle (her rumen), a tight shelter, adequate hay, and a good deep litter, the doe is warm, hardy, adapted to some outside jaunts in deep winter, and capable of surviving many sub-zero nights. She will also be protected against pneumonia. Milk production drops off sharply under these conditions and with this system of hous-ing when deep cold rolls in for real. Mackenzie's milk yield dropped to five pounds daily in that sort of weather, but that was also over an extended two-year lactation. Much of her food intake and energy was being utilized to keep warm.

An alternate method of housing used in these rough climates is the one I outlined first. It is used by people who wish to keep the average production level of the herd as high as possible and who have a pro-fessional breeder's orientation, an outside support income for the herd, or little or no grazing land. They provide an insulated barn, often with the addition of heat in the form of many animals housed together, heat lamps, or a heating appliance or system. The objective is to provide a warm, close atmosphere, considerably milder than the deep cold outside temperatures and harsh weather. Under this sys-tem, the goats are stalled for the winter and not allowed outside for grazing purposes. There are forays to an exercise yard when milder temperatures and better days occur. This approach is usually used for herds of many head or high-production and show herds, animals rep-resenting a costly stock investment, or herds of non-Swiss type where kiddings occur regularly during winter months. Good quality hay with a substantial legume content is requisite because one of the goals of this method is to maintain high yield even through difficult weather. If less goat energy goes toward keeping warm, higher yields can be maintained. This arrangement takes more cash input because it depends upon outlays for insulation, fuel for heat, possible new structures, costly feed and fodders, and labor costs for the hauling and tending. Opting for this type of housing should be warranted by the keeper's seriousness, prior experience, physical situation, and stock investment resources.

If your climate has adverse rainy or deep cold periods, you will need hay to carry your goats through these times. You'll also need storage for it. Loose hay, unless you have been taught to make a

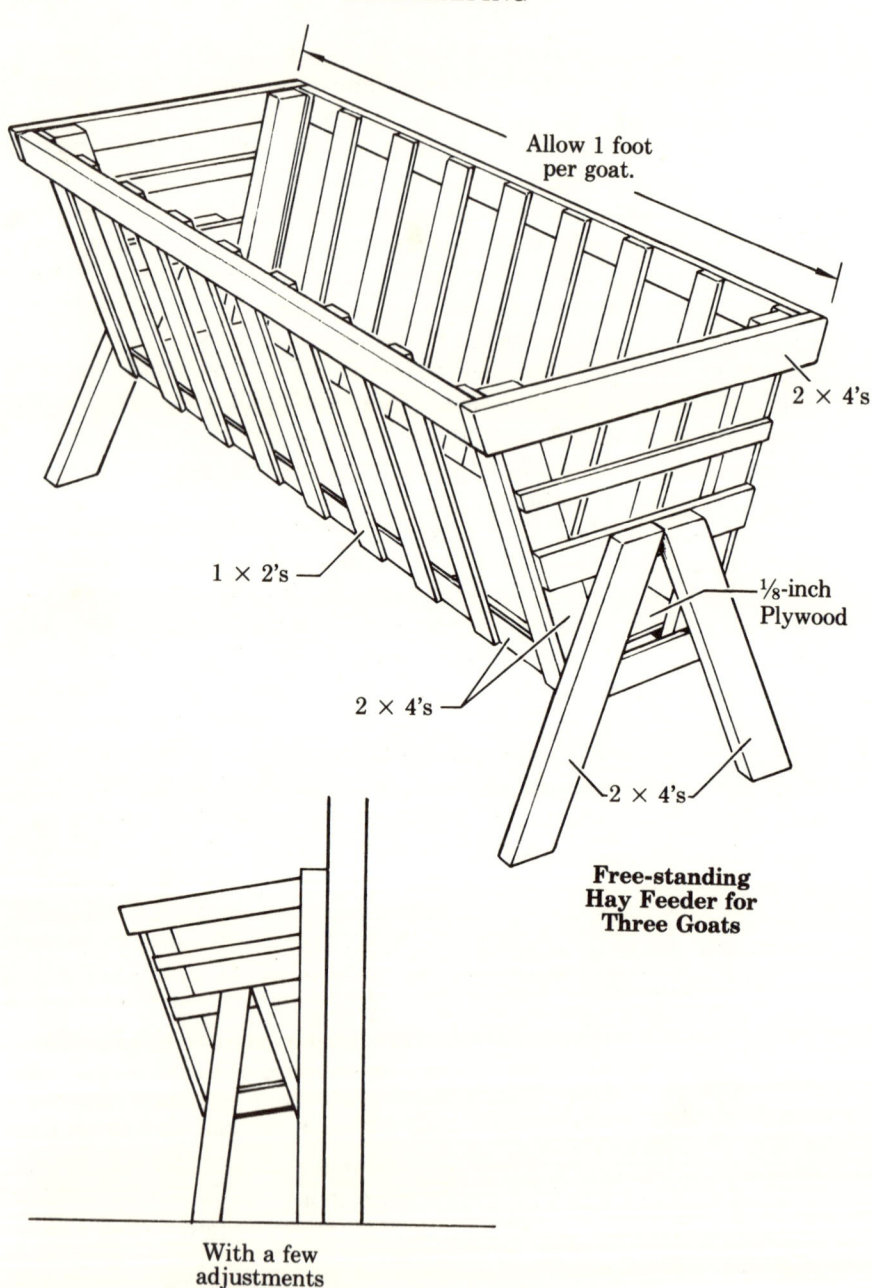

Allow 1 foot
per goat.

2 × 4's

1 × 2's

2 × 4's

⅛-inch
Plywood

2 × 4's

**Free-standing
Hay Feeder for
Three Goats**

With a few
adjustments
this manger can be
wall mounted.

proper outside storage stack[13], is not as compactly stored and conveniently handled as is baled hay. The amount of hay you will need to store depends upon the duration of your local bad weather season, how many head you keep, their ages, sizes, and condition (in-milk, in-kid), and the quality of the hay. If it is good hay, the goats will need fewer bales. Goats throw around great quantities of poor hay by looking for the widely scattered choice tidbits. If you have good arrangements with a nearby hay farmer and he has storage room, you may not have to provide room for storing large quantities of hay in your barn. There is very little farming left in my immediate area, and I have to go a long distance for the hay I buy. We found it more convenient and less expensive after our first winter's experience to lay the entire winter's supply well in advance. In this way, hauling in snowy conditions and high, later-winter prices are both avoided.

You must also provide for grain. Metal containers are an absolute necessity because they are rodent proof, and rats are destructive and menacing. Bringing feed grains onto your land will attract rodents, if they haven't already discovered your treasures. Be prepared, and don't invite newcomers; use metal. If you have a metal-lined bin or chest that can accommodate one- to three-hundred pounds, you are all set. I am using twenty-gallon galvanized trash cans and a large steel tub salvaged from a clothes dryer. When the barn gets crowded with new kids in the spring and we need the room, we put these waterproof, rodentproof, portable metal containers outside and set up temporary kid pens indoors. We have not had any problems with marauders in the grain. If raccoons are a problem near you, you may have to tie down the lids of your storage containers every night.

We store as much as we can physically accommodate to cut down the gasoline expenditures of frequent journeying to feed stores. If your dealer can deliver to you, you needn't allot so much space or concern to grain storage. If you are growing your own feed grain, you will need substantial storage space, but you may already have appropriate buildings and storage bins.

Adjoining or within your goat barn should be your milking place. I use a milk stand with a crossbar locking arrangement. It is in the barn but away from the stalls. Near it are a few shelves that hold hoof trimming tools, a wound dressing, paper towels, and the iodine-based, antibacterial udder-washing solution used at milking time. While the goats are in-kid, I keep cider vinegar on the shelves. Shelving should be minimal; straight surfaces do not provide as good hide-outs for coliform bacteria as do nooks and crannies.

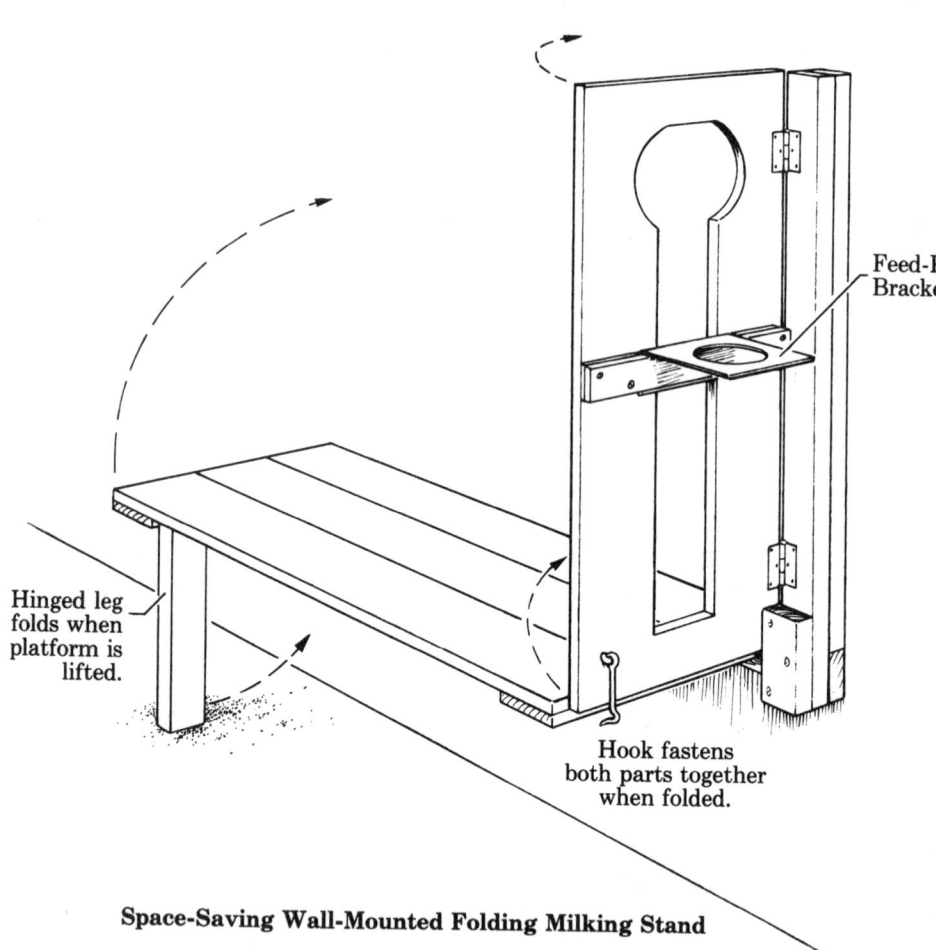

Feed-P
Bracke

Hinged leg
folds when
platform is
lifted.

Hook fastens
both parts together
when folded.

Space-Saving Wall-Mounted Folding Milking Stand

Pasture

The goats' pasture was at some distance from their barn during the first two years of my goatkeeping. This is a poor set-up, and it should be avoided at all costs. Provide direct access from the overnight shelter to the pasture. Much extra labor and cajoling is required to herd goats to and from their pasture when it doesn't adjoin the shelter. Having a distance to traverse also gives them opportunities to drool over and leer at the garden, the *Rosa rugosa,* the fruit trees, and the Jerusalem artichoke plantations. Eventually their curiosity wins and

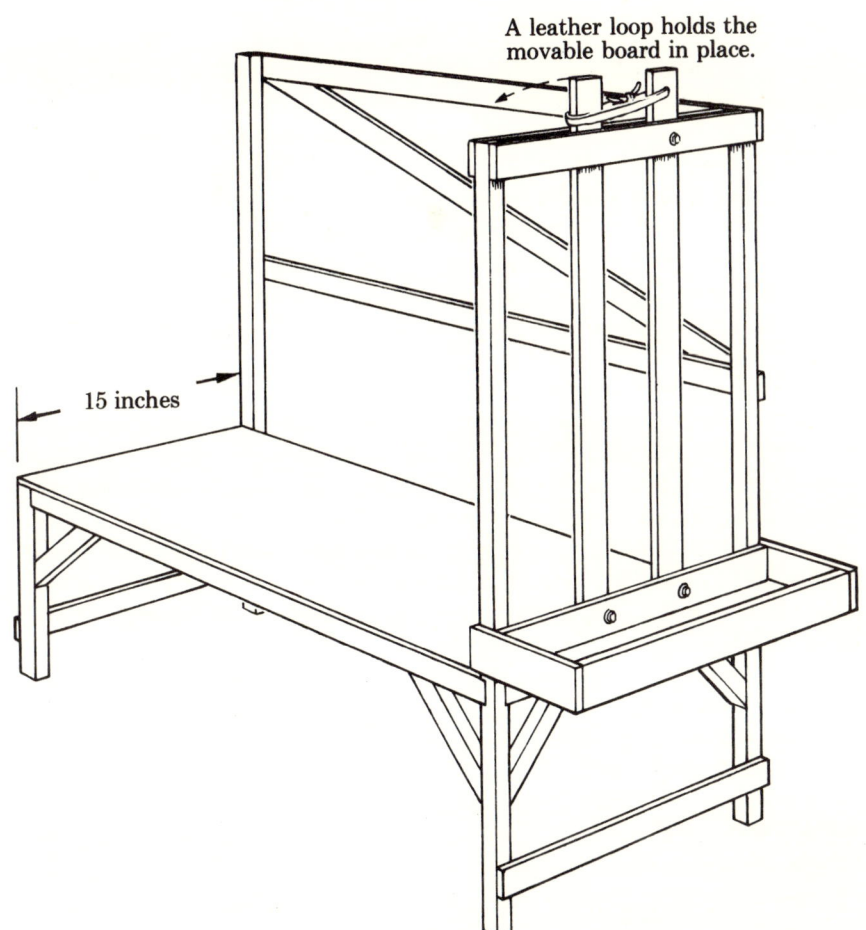

A leather loop holds the movable board in place.

15 inches

Stationary Milking Platform

they won't be herded or led past these forbidden delights. They are very successful at actualizing their food fantasies and will make many sudden rushes at all the cultivated food plantings. They will be quite happy with the variety in the pasture if their daily circuit doesn't include a stroll past the garden on the way from barn to pasture. We took up one bit more of the barn space to change the pasture access arrangements. We mounted an electric fence control unit in the barn and fenced-in a corridor running from the barn door to the pasture.

This three-wire electric fence is more fully treated in chapter nine.

Our pasture is between one-and-a-half and two acres in size. With winter hay and all feed grain purchased, we have luxuriantly grazed four goats on this area. The pasture we have is scrubby grazing, pure and simple. It's never been fertilized, seeded or cultivated. The predominant plants are ryegrasses, yellow sweet clover, white clover, burdock, curly dock, dandelion, chicory, plantain, and a wide assortment of thistles and seasonal late wildflowers such as asters and escaped phloxes. The entire area is bordered with fifty-year-old weeping willows, pines, swamp maples, wild grapes, and some spruce. Sumac was proliferating everywhere and encroaching upon the mature plants but sumac leaves, twigs, bark, and stems turn out to be great goat favorites. Even though our goats prune all summer and strip the bark constantly in the winter, to the point of becoming freckle-faced with resin dots, it is an even contest. I am not sure which force will triumph. Before our eyes we see the perpetual battle of agricultural humanity attempting to push back the primeval forest. The goats diligently clear the scrub but the sumac springs back. At least the goats can keep that acreage from being taken over by wild dense scrub till we can get to cultivating it.

If the pasture looks cropped down, or we want weeds trimmed down elsewhere, or we feel the goats need rotation for worm control purposes, we make use of tethers. Tethering is a nuisance, hard on the goats and the keepers, and dangerous with young or scared stock, or in a vicinity abounding with large dogs.

The goats go out to pasture soon after the morning milking and are out until near dusk from March through November. The exceptions are wet or overly humid, buggy days, such as the days when our local mayflies descend in fierce numbers. The pink-skinned, white-haired Saanen does we have are harassed terribly by insects on hot, sultry, low-pressure days, and head for shelter of their own accord. During the cold, snowy months, the goats are out every day that the weather is tolerable. The grazing consists only of bark, willow tips and accessible evergreens. We supplement with gathered windfall and evergreen prunings we get from our local dump, and of course, hay.

The approach is probably the most viable for the homestead or household goatkeeper, unless there is great interest in showing or breed improvement. Such an interest must be subsidized by an outside income. As one of the owners and managers of a foremost goat-farming dairy has pointed out, very few American breeders derive their entire income from their goat dairying. And David Mackenzie, long ago, distinguished goat breeding as an activity that depended upon outside money. It is not the same as goatkeeping or goat farming, though it shares the same concerns and goals when it is serious and production-oriented.

My choice was also dictated by my conviction that the goat should be an ecologically defensible animal, not a walking grain bin, as Mackenzie scornfully describes some "well kept" milk goats. I was convinced that she could derive a truly coarse, minerally balanced diet from my weed collection and that she could do it by selecting a part of her own food on a reasonably sized pasture (that otherwise would soon revert to scrub), while performing the services of an on-the-hoof composter (as described in all those *Organic Gardening and Farming*® articles I had read), and keeping our acreage trimmed back. At this writing these services have been rendered, and our goats have been hardy, healthy, happy and generous enough to supply our family needs, while requiring only a small cash outlay.

Supply the Essentials

The household keeper can get by with a minimal shelter for the small herd as long as it supplies the essentials adequately. There are so many elaborate plans available that depict detailed housing arrangement that I have chosen to simply refer you to the bibliography if you are interested in building. Get the books out of a library, copy relevant plans and have them on hand for your own guidance. For persons in situations similar to my own, I'd like to present our own construction experiences. Few goatkeepers have the intention of building a true barn for the herd. It is probably even unwise to commit so much time and money to building if you are a novice keeper. Three to five years' minimum experience in goatkeeping should probably precede any elaborate construction project. Plans are valuable, however, for information on how to deal with converting existing structures into adequate goat housing.

We started out calling our barn the goat shed, but after about eighteen months the term "goat barn" worked its way into our vocabularies. The building itself was exceptional as a first housing facility. It was a good, tight, stucco outbuilding, 12 feet by 14 feet, with a peaked roof, and a wooden floor over a concrete slab base, but with a window on each wall and the north-facing door which I have mentioned. Minor roof leak repairs, tightening up the windows and stall construction were all the building needed. Horned goats require separate stalls or separate stanchions during the night and at feeding. The central or communal stall is very convenient, as well as being less expensive to construct, and it provides a social environment, but it is for dehorned goats only. What is sociable contact among dehorned stock is dangerous interaction among horned goats. I still incline to separate-stall housing for household mini-flocks, even when the goats are dehorned, because goats of differing ages and origins discipline or bully the younger or more mild-tempered goats and sometimes manage to successfully exclude them from hay feeding. Even if they inflict no physical damage, they can have an inhibiting and negative effect on the meeker goats. Our does have profited from having their own spaces to retire into overnight and from the separated feeding arrangements. Having separate stalls also does away with the need for separate kidding pens. The does kid in their own stalls.

We made no cash outlay for wood or other construction materials except for 150 board feet of rough-cut lumber at twelve cents a board foot (eighteen dollars) to floor the hayloft. All the other materials — 2-by-4s, siding for partitions, stall doors, hinging and hardware — were scavenged. We have done all the construction ourselves. It is not elegant looking; rather it looks like it was built from the odds and ends that were its source. The barn is a monument to labor-intensive technology as opposed to capital-intensive technology.

Housing a Buck

I'll explain further along in more detail why keeping a buck requires a separate barn, but I guess most people know. HE STINKS. There is no getting away from the tangible reality that sexually mature bucks reek. The odor is offensive and persistent, clinging to anyone and anything — including the milk does and the goatkeepers. The buck odor passes into the milk of the does if the does are allowed to hang out with a good, potent, stinking buck. The consequences of housing bucks with does have given rise to part of goat milk's strong, negative reputation and general bad press. I had never planned to keep a buck. The vision of trying to persuade 175 pounds of masculine sexual aggression to my way of thinking some bright October morn when he was bound to have other things in mind made me discount the possibility of keeping bucks right at the start of my goat dairying. I had already seen an experienced, hefty male goatkeeper knocked flat by his favorite buck when the animal was expressing affectionate delight at the man's visit. Even Mackenzie could haul me around the yard any time she took a mind to. However, new realities define new directions. With the advent of the fuel crisis all my antibuck feelings evaporated. Breed fees for the fall of 1973 came to seventy dollars, exclusive of gasoline costs and vehicle wear and tear. I was also trying to be scrupulous about breeding-up, even though all my goats were scrubs. I felt every keeper had an obligation to breed-up, not so much because of the higher price the kids would command, but because I have a passion to see milk productivity among homestead goats rise. Such passions necessitated journeying over two counties to breed-up the does to purebred bucks with milky pedigrees. Since all the does don't oblige their keeper and come into season on the same day, come fall one starts covering a lot of miles, burning a lot of fuel, and having a lot of misses and returns. One *can* decide to breed one's does at the nearest buck pen, but that was not the path I was on.

I wanted to learn about bucks "some day." The day had arrived, courtesy of Exxon, Shell, Mobil, and other forces usually far from the thoughts of the pastoral goatkeeper. Instead of first rushing out to buy a buck, we had to put up a buck barn. This was more than a two-person endeavor, so we had an old-style mini-barn raising. We threw a Leo birthday party and asked friends to bring food and hammers. Up went the skeleton of a 12-foot-by-7-foot shed-style addition to the doe barn. Almost all the materials for the project were scavenged: rolled roofing, tar, timbers for the roof, log poles for the walls, siding, door, and windows. Two exceptions were made. We purchased a single sheet of ⅜-inch plywood sheathing to fill in some gaps in the siding, and we bought some steel reinforcing rods to tie the sills into the ground. The total cash outlay for these materials and a case of beer came to fifteen dollars.

The site provided half the structural support necessary since the addition is tied into the original building. The site is at the top of a rise and has unusually good drainage, so that we are able to utilize a log-cabin type of foundation with shallow, rock-filled trenches. What we built is basically a dry, windproof lean-to for our purebred, pedigreed French Alpine buck with enough windows to allow adequate ventilation for the buck. We hope that the ventilation will keep both the buck and his quarters from reaching the outer limits of raunchy "buckness." We now have added a small flock of chickens. "Hardly smells at all," some folks say. "Must be the buck-o and the chickens canceling each other out," says I to myself.

CHAPTER 9

MANAGEMENT AND CONTROL

Control is fundamental to management. Nothing can be directed or guided if it is out of control. Control is also essential if the homesteader is to feel gratified and rewarded by goat dairying. Are there days when your nearest and dearest neighbor comes to complain that your doe has leaped the garden fence and is gobbling up a whole winter's food supply, finishing up with a rose bush dessert? Or afternoons when your heaviest producer is bullying all the other goats and kids around the pasture? Or mornings when the youngest and most thriving of the apricot trees and comfrey plantings are decimated in a ten-minute spree? Are there constant fence-mending walks, scenes at the milkstand, sounds of outraged and vociferous goats coming from the barn? These are signs that the goat enterprise is slipping from the keepers' hands and into the control of the goats.

Locks, Fences and Goats

Goats are notorious for their fence-jumping and latch-opening talents and their true catholicity of appetite. As a matter of verifiable fact, goats possess these capabilities and love to display their talents to the fullest. It is up to the keeper to see that the goats' appetites are well met, their fences adequate, their latches secure. In this way their natural abilities and tendencies remain prankish and do not become destructive patterns, vicious and entrenched habits.

There are no goatproof latches, a fact attested to by beginners and veterans alike. I have found double latching the most secure arrangement of the many I have tried. Hook-and-eye closures are suitable for young kids, but I always use two on each door. The spring-lock type is best, even for little ones. Goats can and will crane out over their own stall doors and open a neighbor's latch or hook and eye. After viewing this seemingly impossible feat several times, I learned to place a lower latch on every door. For the doors of mature goats I use latches with sliding bolts. I have had numerous door-battering events since I have had several older does with full horns. The best arrangement I have come up with for these sometimes grouchy older ones, short of tethering them inside their stalls, is to hang the stall door so that it does not close flush with the surrounding verticals and to fasten it loosely with two lengths of chain, top and bottom, with double-ended clips attached for quick opening and closing. A fully horned doe can be very assertive when she comes into estrus or when she enters a new herd, always challenging for the leadership position and throwing her weight around. When such a doe gets into a pushy mood she'll use her horns against the chained door. It will yield a bit, but not shatter apart upon impact.

Once a goat learns that she has the power to open her stall door, she will repeat the trick over and over. Goats are great ones for grasping the implications of personal power. There do seem to be elements of learned and earned delight and gaming in goat psychology. Goats will continue to open latches and escape their stalls if new, tighter, arrangements are not made immediately. They will continue this performance, even though there are no specific gains or rewards, such as grain-bin access or another prize, achieved in the escape. Perhaps they are just delighted with the keeper's consternation at finding everything in the barn out of order — doors open, an animal or two in the aisle, upset pans and buckets, or — the old favorite at our place — the roll of paper toweling for udder cleansing pulled around and unrolled throughout the barn. Pleasure in disarray is a goat trademark. It must be clear by now that escapist antics make for much added labor in the barn. I have learned to respect the dairy goat's intelligence and potential for rapid learning and have started my control program with the individual stall.

Patterns

After tight latches the next element in the control of the domestic goat is pattern. Goats respond beautifully to repetition and patterned activity. Routinize all barn activity. Feed and milk all the animals in the same order, day and night. Water them in the same order if water is not available in each stall. Let them out into the pasture in the same order; bring them back, and stall them in the same order. At times these patterns may have to be adjusted, such as when a newly fresh doe joins the milkstand brigade or when a milker has to be isolated and milked last for udder health reasons. If possible, make these changes slowly. If such a break in the order of things has to come quickly, offer the goat that is to be fed or milked out of routine some part of her grain ration or a fresh vegetable treat (I use the pulp left over from juicing carrots), while she waits her turn. These compensating treats forestall the indignant behavior that arises from a quick change. The goat's internal clock seems to be very sensitive. If the keeper shows up at a thirteen- or fourteen-hour interval when the goats are well accustomed to twelve-hour intervals, there may be some antic behavior ahead. If you know that you have to be away and will be forced into a longer-than-usual interval, placate the goats' time clocks by an extra hay feeding in the late afternoon or by gradually lengthening the feeding interval by half an hour each day for a few days prior to the day on which you must be away. Just suddenly appearing an hour or two later than normal will almost guarantee attempts at break-out or irascible behavior on the milkstand.

Patterning also works extremely well with goats' notorious fence leaping. It is important to establish an acceptable browsing circuit for the herd. There will be many fewer escape incidents if this is done from the start. It may sound somewhat silly in this far-from-pastoral age, but I can recommend that the keeper lead the herd, even if it is a herd consisting of only two, into the delineated pasture. Lead the goats on a browsing-cruise to set up a grazing pattern. This is most easily done by clipping the eldest doe to a lead and taking her for a prolonged stroll of a few hours' duration, starting from the barn and making the rounds of the pasture area, then slowly returning to the barn. The younger or less experienced doe will follow the two leaders about. Any older doe with real foraging experience will perform this service for the keeper and the herd, but I have found it very helpful to delineate the limits of the acceptable territory in this way first, and then to relinquish the leadership role to the proper herd animal. This procedure is a real necessity when dealing with young, inexperienced stock. It is only after the goats have accepted a demarcated circuit that they will truly respect fencing.

Fencing Pastures

Once the herd has learned respect for fences the fence can be counted upon to control herd movements. The reverse order does not seem to work; one cannot simply bank upon the dairy goat's awe of fences to make her stay put. Of course, if one can afford to provide five-foot-high, small-webbed steel or galvanized fencing around an acre or two of pasture, teaching the goats a preordained circuit may not be necessary. I have read of people who fenced acres of pasture with high, chain-link fencing, but this alternative is financially unfeasible for most keepers. I have seen people fence with a combination of low stock fencing and barbed wire strands, a poor arrangement for goats. Even does in milk seem to behave as perpetual doelings, thinking nothing of occasionally attempting a four-or-five-foot fence jump. When there is barbed wire involved there is always a possibility of a disastrous udder accident. If you plan to fence with thirty-six-inch or forty-eight-inch stock fencing, use a single strand of electrified wire at the top, rather than barbed wire. If an existent barbed wire fence is in good condition and you choose to avail yourself of it, make sure all animals are innoculated against tetanus.

I have had good control with electric fencing by utilizing a *three*-strand arrangement of wire. The goats should be patterned to the defined browsing circuit as outlined above, and then conditioned to the fence. Do not back off on the conditioning process if you plan to rely on electric fencing. Conditioning is based on luring your goats to the fence with some preferred treat and then letting them learn firsthand about the fence's shocking character. In the beginning it may seem a kindness to the goats to omit such specific conditioning sessions, but this may mean headaches for the keeper later on. I have seen goats respect electric fencing even when no current is being run through the wires because their original introduction was sufficient for them to learn to respect the fence. Conversely, goats will escape even a live wire fence because no original fear or distaste was impressed upon them.

One major drawback of electric fencing is that it either requires a site that has electric power run into it or it must be powered by batteries. Additionally, that power is drawn upon constantly for seven to ten hours a day almost all year. Nevertheless, the electrical usage is very minimal, about the same as constant use of one 75- or

100-watt light bulb. I compensate for using electricity in fencing by using less of it in other areas of my life; I also often turn the fence off once a stable group of animals has been conditioned. This is not a viable control method when the fence is a new experience for the herd, but I have done it successfully for months on end when dealing with animals used to the fence. If you own a buck, don't turn off the fence during heat season. Don't turn the fence off at times of the year when pasturage is minimal either. These times are usually the late fall and periods of early spring or midsummer in a dry season. These are times when browsing animals are most likely to want greener pastures, the proverbial times when the grass is always greener on the other side of the fence — which means the goats are longingly looking at the garden, the fruit trees, the rose bushes, and the flower beds. Even well-trained goats will risk the awesome fence or test it a lot if their usual pasture is short on grass.

Fence Maintenance

Electric fence maintenance is constant, mainly involving ongoing walks along the fence line. The keeper has to keep down the weedy growth under the lowest strand of wire. I use a hand scythe or golf club cutter for this. The keeper must also check to see that all the wires and the insulators that hold them are in place. Storms and passing animals can take wires down, or even knock over a fence post at times. It is useful to mulch areas along the fence line that seem to grow back especially quickly or luxuriantly. Six to eight inches of black weatherproofing paper, held down by rocks, works quite effectively to smother growth right under the line of the fence wires. Laying out such a mulch reduces the grass-cutting maintenance time by many hours. Gates are placed at strategic points along the fence for the entry and exit convenience of the keepers. Gates are formed by using ready-made handles designed to allow the wires to run through them. The handle has a hook that fastens onto a wire eye attached to the fence post. As these hooks age and are exposed to snow, rain and humidity, they will rust. If enough rust builds up on the surface of the hook, it will prevent the transmission of electric current. It is helpful to carry a piece of emery cloth on fence-surveying walks. Check all gates and connections or splices for rust accumulation and sand it away where it has occurred.

All fencing must be checked and restored after the inroads of northern winters. The snow and ice drag down wires and heave posts not sunk below the frost line. It is hard to restore fencing until the weather settles down in the spring and the ground is thoroughly thawed. This transition period of sagging fences or wires can endure for a week or two and may present herd control problems. If only a few goats are involved, the keeper can resort to tethering them outside for a few hours a day or accompanying the animals on short exercise strolls. For four or more animals, it might work better to fence off a small temporary exercise yard by stringing electric wire to available trees and the standing posts.

Even though one must be constantly vigilant about the maintenance of electric fencing, it seems to require less labor than other types of fencing and certainly is cheaper to install. Fewer support posts are needed than in other fencing arrangements because the light strands of wire need not be supported by frequent, closely spaced posts. The costs of the control unit, insulators, and wire combined come in well under any other type of metal-mesh or weatherized webbed fencing materials.

The control unit can be battery operated or use 110-volt AC current. It is important to use a unit that is Underwriters' Laboratories approved so that no accidents befall people or animals. The unit itself should be mounted where it can be clean and dry; roof it over if it is mounted outside. Connect one terminal to the fence wiring. Connect the other to a ground rod or pipe extending down into permanently moist soil. Connect a lightning arrester between the fence line wire and the ground line. The units I have used emit a constant bleeping tone and flash a small light when operating optimally. Failure of these signals usually indicates that the fence is grounding out somewhere along the line — the keeper's sign that a fence inspection is overdue. In most instances, brush and weed growth touching the lower wire will make the fence ineffective, despite manufacturers' claims that the unit keeps the weeds back by electrical-shock weeding. The keeper must do the weeding.

Set and brace corner posts well. There are now cornering insulators on the market that are convenient and worth searching out. The fence will probably have to run over some slopes and uneven stretches of ground. I have found that eliminating the lower wire and placing the posts a little closer together is the easiest way of handling slopes. Splicing is best achieved through what is called the Western Union splice, a splice that ensures good electrical contact. Stretch wires tight so that there is little sagging. The wires are supported by fence posts and attached to the posts on insulators, once brittle porcelain but now primarily plastic (still rather brittle after a winter or two). Too much

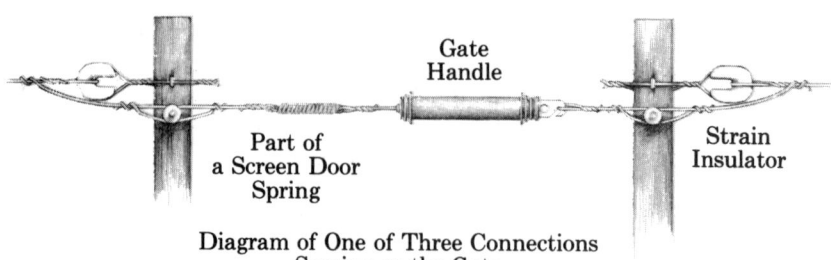

Gate
Handle

Part of
a Screen Door
Spring

Strain
Insulator

Diagram of One of Three Connections
Serving as the Gate

How to Handle
a Rise in Terrain

Strain Insulator

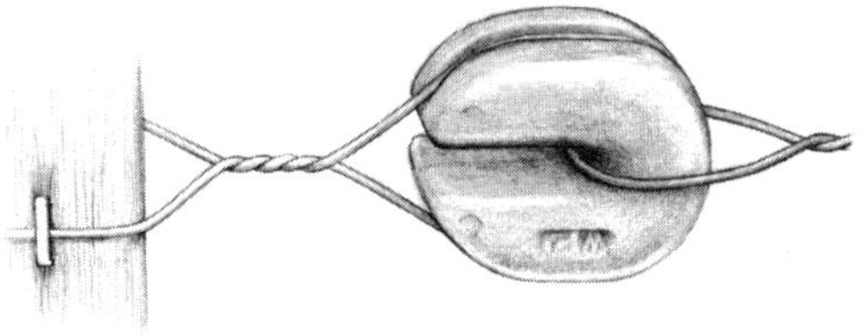

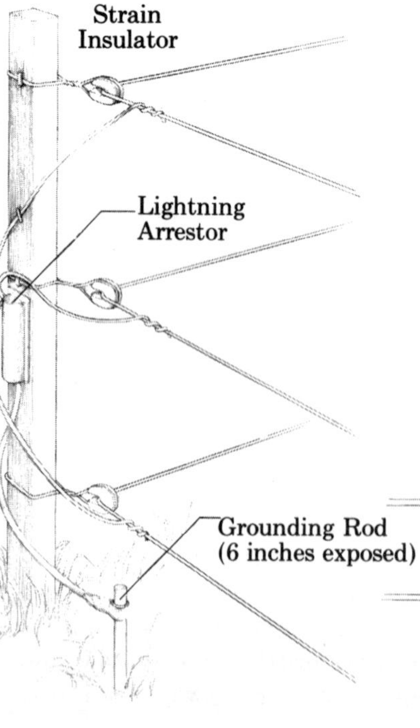

Strain
Insulator

Lightning
Arrestor

Line Post

Porcelain
Insulator

Double-Headed
Nail with a
Plastic Washer

Predetermined
Wire Height

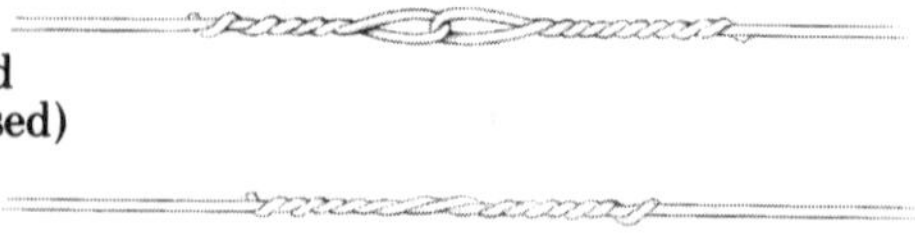

Grounding Rod
(6 inches exposed)

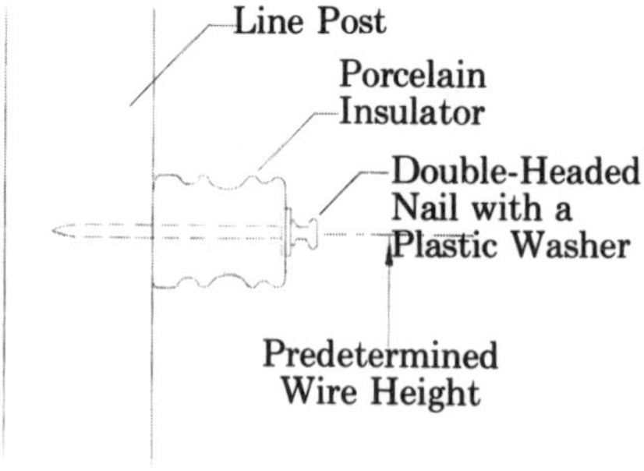

Two Ways to Splice
Electric Fence Wire

stress on the insulators will cause them to snap. Wire should be taut enough not to sag between posts, but not so tight that the insulators are pulled away from the posts.

The easiest way I have found to set fence posts is to dig a starter hole with a heavy-duty wrecking bar, and then to pound in the post with a sledge hammer. I only use cedar or locust trimmed and pointed posts; it really cuts the labor enormously. It is much more efficient, effective, and enjoyable for two people to fence than for one to take it on as a solitary project. Setting the posts, stringing the wire, and mounting the unit for a one-and-a-half-or-two-acre pasture can be achieved by two comparatively inexperienced people putting in twenty to thirty hours of work together. The maintenance can easily be handled by one person. In 1975, the cost of fencing my almost-two-acre pasture was seventy-five dollars, inclusive of unit, insulators, forty cedar posts at seventy-five cents each, and wire. I utilized trees where possible. Nail up a support cleat to the tree and then nail the insulators to the cleat. If you don't, the tree will grow around and engulf the insulators, causing the fence to ground out in those spots.

When repairing the fence, remember to turn current off at the main switch. Avoid handling connected (live) wires or equipment when the ground underfoot or your hands are wet. Many control units do not have an on/off switch, so pull the plug (not the cord) of the appliance to disconnect the unit. These sound like elementary reminders. They are. But when mending comes up unexpectedly in the course of other business, people often overlook them. If you touch the fence once when it is live, you won't do it again.

Tethers

Keepers who do not have extensive pasture for their goats may have to rely upon tethers or small yarded enclosures. Goats kept in these circumstances do present more control and management problems. Tethering is a frustrating experience for keeper and herd animal alike. If it becomes a must, it is less frustrating and dangerous to employ chains, clips, and swivel devices so that the animal has as much freedom of movement as can be achieved within narrow limits. Yarded goats will mind their fences and accept limitations if the keepers are diligent and attentive to their feeding and watering needs. Cut as much fresh greenery as possible for goats in tightly limited environments. Parasite infection can become a larger problem than control in these situations. Readers are referred to Jeffrey Williams' excellent and detailed article on "Internal Parasitism" in the May 1977

Dairy Goat Journal for a technical understanding of how best to cope with this problem in goats kept in enclosed spaces.

Space, control, and herd health are inextricably linked and serve as an excellent example of the intertwinings of the art of management and all its subsidiary activities. In the context of dairying, management is the whole art of handling, taking charge of, and directing the entire enterprise. Books can be written on this topic alone, and indeed they have been. Thus far, much of this book has had to do with management advice, technique, and practice. The rearing and keeping of truly productive livestock is grounded in care and responsible management. In closing this chapter, I wish to summarize and reemphasize some management points:

— Move slowly if this is a first livestock attempt.
— Watch out for enthusiastic overexpansion.
— Review purposes, climate, and resources.
— Choose numbers, breed-type, system of maintenance (i.e., yarded or pastured, cold barn or warm environment).
— Establish goals and delineate limits to stick by (i.e., show orientation or stock improvement or household production, i.e., vocational or avocational, income yielding or not).

Management of productive livestock is a complex art and the Greeks knew what they were talking about when they said, "life is short and art is long."

CHAPTER 10

WELL-BEING AND ITS DIVERGENCES

A well cared for, reasonably nourished milk goat that lives in a good, tight shelter and has access to outside browse is hardy and healthy almost all the time. Her pregnancies are usually calm, fruitful, and nondebilitating. The focus on illness commonly found in the disease and health chapters of many goat books runs directly counter to the fundamental soundness of the dairy goat's constitution and general capacity for well-being. Problems definitely arise over a long-run course of goatkeeping, particularly when herd numbers increase rapidly beyond the resources of the keeper and the land, or when there is close, inside confinement. The basic reasons for including a health chapter in this book have little to do with the nature of the goat *per se*, but more to do with the current realities of novice livestock keeping and rural veterinary service.

Most of the newcomers I have met keeping goats today are people with little or no farm background. They are not habituated to careful observation of livestock. Usually their only prior experience keeping animals has been with cats and dogs in confined urban environments. There, when problems developed they called a veterinarian and took the animal to an office. This behavior doesn't transfer well to goat-keeping, so the novice is often left high and dry. Trouble escalates quickly into a major difficulty when you are unaware of what is supposed to be noticed routinely. Problems are compounded by the unavailability of veterinary care. Many homesteaders have urgently needed professional assistance but found every veteriarian they called was dealing exclusively with small animals and made no livestock calls whatsoever. Others have found themselves miles from the services of a livestock-oriented veterinarian. Some folks never investigate the veterinary resources of their area till their first crises are upon them. There are seven veterinarians within a fifteen-mile radius of my farmstead, but the nearest livestock vet is twenty-five miles away and has an already heavy schedule of commitments. In such circumstances it is not a simple matter just to ring up a vet because trouble has arisen.

Given current realities, it is wise to check out your resources as soon as you get two goats. Consult your local cooperative extension and Society for the Prevention of Cruelty to Animals, and any dairy farmers in your vicinity. Make a list of veterinaries who are handling livestock calls in your area. Call each one of these vets and briefly describe your situation. Tell him that you are keeping milk goats, how many of them, and how old they are, and give him other pertinent data. It often turns out that individual veterinarians have had little experience with goats, but sometimes you will find one who has kept goats or has seen a lot of dairy goat herds professionally. If so, you're in luck. Find out how far it is to the vet's office. Goats can easily be transported, even in the back seat of a car in many instances, but it is not advisable to drive them around late into the gestation period or in severe weather.

Arrange a call, based on mutual convenience, to have your animals tested for TB and brucellosis. This service is often free to dairy herd owners if you have a number of animals in milk. If you are keeping only two, you will probably have to bear the costs of the tests yourself. Keep costs low by bringing your goats to the vet. The same is true for routine immunization, such as that for clostridial diseases. If you find a vet who has a route of calls already scheduled for your area, his or her on-farm call may be reasonably priced. There are also veterinarians who have countywide responsibility to do dairy herd health tests. This is easily checked out prior to an emergency, and having the facts at your fingertips gives you an advantage when emergencies do arise. It may also be a good idea to query the vets you speak with about their experience with dairy goats.

A straightforward listing of ailments in alphabetical order seems to do little for the keeper, for it gets cursorily read and consulted only in times of dismay and confusion. This chapter seeks to read as a narrative rather than an encyclopedia of worry and disease. It is organized in categorical groupings of what I perceive to be the most prevalent contemporary problems in order of threat: parasites, abscesses, diseases of overeating or underfeeding, pregnancy and pregnancy stress-related conditions, and milk production difficulties. Not every possible goat problem is treated in this chapter and some old, widespread problems — such as pneumonia, which has not disappeared but does not seem to be the great killer it was in the not-distant past — are scantily handled. My focus is on acquainting the newcomer with basic trouble signs to keep in mind at all times, and teaching how to spot trouble early and how to make sound decisions about calling in professional help.

Know your animals: that is the key to goat well-being. Though it sounds simple, knowing goats is actually quite a complex art; you must be well acquainted with individual animals, know their needs, habits, and idiosyncracies, and have in mind a general standard of health against which they must constantly be measured.

A goat in good health is alert, vigorous, and animated; she approaches life with interest, curiosity, and intelligence. In translating these qualities into action, she responds to her keeper's approach or call, not necessarily as would a dog, but with an active interest in arrivals and departures. She usually moves with her herd or companion with visible eagerness to go out to pasture. She vocalizes, jumps about, assumes alert stances in response to her environment. She carries her weight on all four legs, planting her feet squarely as she goes, when not running, leaping or gamboling. Her appetite is active without being voracious. Her skin is smooth everywhere, pliable, and free of lumps, bumps, or puffinesses. Her hair is sleek and soft, whether long or short. She passes urine that is a clear, yellowish color, and feces that are neat, dry, dark pellets. The mucous membrane tissues around her eyes, nostrils, vulva, and mouth are pink-colored. The groundwork of preventative care is laid by observing your animals every day with a view toward their general well-being. Use the above description as a mental checklist. After awhile it becomes automatic.

A goat with a developing problem varies in some way from these norms. The first signs I have always seen are:

    Changes in appetite: loss of interest in food, erratic appetite, voracity.

    Changes in carriage or animation: listlessness or dullness, hanging back from the herd, slow or unsteady movement.

    Changes in fecal matter: clumpy, mucusy pellets, diarrhea, blood in the stool, or changes in the urine such as blood in the urine.

Not all these signs appear at once, though combinations, such as the appearance of an erratic or voracious feeding pattern accompanied by a change in fecal matter, are common. (Also read the section of this chapter on internal parasites.)

Normal body temperature for a goat is between 102° and 103°F. (between 38.9° and 39.4°C.). The pulse should be between 70 and 80 and the respirations between twenty-two and twenty-six per minute. A high temperature can shoot the pulse to over 120. Take your animal's temperature now and then to get both you and the goat accustomed to the procedure before crises evolve. Clip your doe up on a short tether or have a friend hold and restrain her, and approach her talking familiarly. Lubricate the thermometer and insert it slowly into the rectum. Take advantage of the moment when she pulls her tail up, or gently slide it out of the way with your hand. If you grab her tail and pull it out of the way, she can get pretty resistant and excited. Half a minute's placement will yield a reading. Check for the pulse rate by feeling the artery high up on the inside of the thigh. Respirations are checked by watching the flank. Going through these checks before troubles come helps the novice understand what is happening when one, or several, of the major change signs occur, and provides additional information that may be needed to make a diagnosis or a decision to call in professional help.

Internal Parasites

All three of the major indicators of change are involved in the most prevalent contemporary threat to goat well-being, parasitism. Internal parasites seem to me to be the most common health problem in goatkeeping today, with roundworms of the intestinal tract and stomach and coccidia leading the pack. Even the most consistently well, trouble-free herd will suffer what is known as "spring rise," a great jump in the numbers of eggs of the roundworm class of parasites in the early spring, or right after the kidding season peak. Symptoms of large numbers of digestive tract roundworms are appetite oddities which can take the form of extreme voraciousness or lack of interest in food; any fecal variation from the normal tight, dark pellets; weight loss; or paleness of the lips, the tongue, or the mucous membranes around the eyes. There may be a soft lump under the jaw.

If you see signs of parasite infection, take a fecal sampling into your vet for testing. Many small-animal vets will be willing to do this kind of testing for you if you are distant from a livestock veterinarian. If it turns out positive, obtain and administer an appropriate worming compound (anthelmintic). These compounds come in various forms: boluses (a large thick pill), liquified, and pelletized with alfalfa or something else attractive to goat palates. The pellets are fine but expensive. I have found it easiest to use boluses, and either pulverize them and mix the powder in the grain, or give them directly by mouth. You'll need an extra hand for doing the latter. Have an intrepid friend hold the goat's mouth open and restrain the goat while you get the bolus into the animal's mouth. Goats have a great capacity for lulling you into thinking they have swallowed and ingested boluses. They then drop the pill out of the corner of the mouth. If this happens, give the bolus again and hold your hands over your animal's mouth and muzzle. She'll give up in a couple of minutes and down the thing. Test the feces again in two to three weeks to check whether anything new has hatched.

If the first test results came out negative, but you continue to see symptoms of parasitism, you may have to trust your own observation. Testing procedures are based on egg counts and sometimes maturing populations of worms may produce severe infestation symptoms in your does but the egg output may be at a minimal point and so fecal exam results are negative. In cases where the egg count is low, but the doe is really infested, it may be wisest to go ahead and worm her. In these instances milk production will often rise, feces will normalize, and feeding patterns regularize. Usually, a semiannual fecal sampling of the barn population will reveal whether or not to worm and when to do so, but results of tests must be related to the keeper's scrutiny of the herd. In my region, the temperate Northeast, I would recommend April to May and October to November as the time to double check for internal parasites via laboratory examination.

All books recommend pasture rotation as a preventative measure because eggs are passed out with the goat's droppings onto pastureland. Within a short time, these eggs pass through two larval stages that can survive on moisture and bacteria. The second stage larvae molt to form the third, infective stage. These larvae do not feed but wait in ambush on pasture plants until ingested by grazing goats. The larvae then affix themselves to the lining of the goat's digestive tract and mature into adult worms, copulate, and produce new eggs.

Goats can and will tolerate a diverse population and sometimes sizeable numbers of digestive tract worms without major stress. It is probably impossible and unwise to strive for the completely wormless goat. However, when *Trichostrongylus* are in the intestines and *Haemonchus* or *Ostertagia* are in the stomach lining, milk yield will drop, and loose stools, general loss of well-being or "unthrifty condition" will develop. Anemia is a possibility if the infestation goes unchecked. In addition, a species of roundworm called *Nematodirus* has eggs that have been found to be extremely resistant to destruction by cold, sun, or drought. The eggs remain dormant for lengthy time periods, hatch out suddenly and can be taken up in great numbers into the digestive tracts of vulnerable young kids. Sudden and acute diarrhea is the major symptom. Prevention consists of not running the kids on the same pasture or yard areas year after year and of administering a dose of a powerful anthelmintic at six weeks if necessary.

Other internal parasites of dairy goats do exist, particularly liver flukes and lungworms. Again, herds being routinely checked twice yearly via fecal samplings and closely observed daily by their owners rarely have unchecked infestations. Liver flukes also need special conditions: wet, mud, and the presence of the snail, *Limnea truncatulus*, to develop into the infective larval stage that can be taken up by goats. Listlessness, dullness, loosened stools, and abdominal distention warrant veterinary consultation.

Lungworms in goats are geographically more widespread than liver flukes. Some lungworms have little clinical consequence. Other types can create severe respiratory difficulties. Constant coughing, poor condition, and nasal discharges are the usual symptoms. Prevention involves avoidance of wet pastures. Treatment is currently based on tetramisole.

It is important to be aware that many small keepers simply do not have enough land for rotating pastures, that small fenced-in areas get grazed down intensively, and that small herds of two to four goats will not venture out very far from their shelters and human friends even when a new pasture is fenced and opened up to them. Contamination by larvae of many types of worms and by protozoa will occur in the small feeding area near the shelter. Williams feels that the "oft-cited methods of pasture spelling and rotation have little application to goat husbandry and in fact have very limited value in sheep and cattle parasite control. Contamination in and around feed troughs is a major focus of parasite transmission."[18] His suggestions are to recognize the dams and other adult animals as the carriers, to become aware of the dynamics of worm population growth over a year's time, consistently to check fecal samples and look for symptoms, to keep feeding areas and equipment sanitary, and to use anthelmintics when vigilance alone does not suffice. In this way only two or three anthelmintic dosings need be given over the year at the more critical junctures, rather than routine dosings every four to six weeks. Milk must be withheld from human consumption for ninety-six hours, or eight milkings, when using the more commonly recommended wormers. Thiabendazole and tetramisol (not currently FDA-approved for goats but being utilized by many vets), are the compounds most used at the present time. Because the milk is unfit for people's use after a doe has been wormed, overly frequent use of these compounds makes serious economic inroads on the household milk supply, and monthly worming is probably not even a healthy regimen for the goats.

Protozoan parasites, *Coccidiosis* Coccidian organisms of the genus *Eimeria* attack the digestive tract of goats, infesting the small and large intestines. The exact *Eimeria* culprit in goats has not yet been isolated. Currently, coccidian infection is a great problem, particularly because it can suddenly strike and kill young stock. Not much is known about coccidiosis in goats, and it is "not possible to use drugs to eliminate coccidial infections,"[19] although drug therapy is available to *control* infections. Traditionally, sulfonamides have been used, but they are now superseded by amprolium (also not currently approved for dairy goats by the FDA, but being prescribed by vets).

The gut of the problem is in the omnipresence of the eggs, called oocysts and in the capacity of the oocyst to survive environmental extremes of cold, heat, or dryness. There is a "rapid development of infectivity in oocysts leading to massive contamination around food and water troughs and restricted grazing areas."[20] Once oocysts are ingested they travel to the intestines, hatch or "sporulate" in a few days, and produce new hosts of protozoan parasites which attack intestinal walls and destroy the mucosal cells. Each of these new parasites can then produce eggs that are shed in the feces while intestinal attack and proliferation continue. Acute cases are common in young kids during which appetite drops away, abdominal pain occurs, and the kid dies in a day or two. Older animals can suffer less severe attacks, with signs being primarily loss of appetite and loosened feces often with signs of blood. Drug treatments will be necessary but are not the entire answer.

If you think coccidiosis has appeared in your barn, take fecal samples from all the goats and bring them in for laboratory analysis. Don't wait! Loss of young stock can be rapid and decimating. If these organisms are found in sizeable numbers, obtain amprolium or one of the newer sulfa compounds from your vet and follow dosage and instructions. Treatment can run from five days to three weeks. These compounds are usually administered in water. Offering fresh, clean drinking water to your animals frequently at this time is of primary importance. Flushing the kidneys is absolutely necessary in order to avoid kidney stress. Certain of the sulfa compounds require witholding the milk from human use for lengthy time periods.

If you have no access to a veterinary diagnosis and loose, bloody stools are rife in the barn and the goats are off feed and miserable, suspect coccidiosis before you sustain losses. You may have to resort to over-the-counter sulfa-based compounds such as Sulmet or Sulfanox. Follow sheep-scaled dosages and provide water in abundance. Relapses at two-week intervals are possible, so be observant.[21]

The most important thing is to limit the number of oocysts to which the young kids may be exposed. Keep kids separate from animals with a past history of coccidiosis. Feed stock from separate, new pans and troughs. If obtaining new equipment is not possible, disinfect the old with a strong solution based on chlorine bleach and then expose it to the sun for an extended time period. Don't raise this year's kids in the pens of last year's kids if there was a coccidiosis outbreak. Oocysts can survive for many, many months. Some people treat new kids with low-level doses of a coccidiostat in their bottles when new pens or yards are an impossibilty.[22]

Wear rubber slip-on galoshes when visiting other folks' barns or when working in or passing through infested grazing areas. Hose down and sun your galoshes regularly. Sunning and desiccation can aid in limiting oocyst survival. By all means, avoid developing crowded, wet conditions in your stock shelters, yards, and pastures.

My feelings about the complex parasite question are oriented to an ounce of prevention being worth a pound of cure. Overaccumulation of animals undoubtedly helps parasite problems proliferate; avoid it in the barn and on pasture. It used to be a rule of thumb that a goat could be supported on a quarter acre, but in reality land quality, type of vegetation and available shelter make for much variation in the rule. Some experienced persons can tend ten goats easily and well, and some people are swamped by caring for three or four. The threshold of overcrowding is different from herd to herd. The goal should be control of parasites rather than eradication. Some of these parasites may even function in some ways to maintain balance in the goat's inner ecology. Aiming for a parasite-free goat is probably a waste of energy that could be more fruitfully channeled into sanitation or housing improvement. However, maintaining control through balance prevents losses to parasitism. Don't depend entirely on all-chemical or on all-natural control. Total dependence on chemical anthelmintics can produce animals with little capacity for self-immunizing responses and with lowered resistance and chronic susceptibility to reinfestation. Total dependence on garlic, rue, wormwood, and other natural vermifuges is unfeasible in the contrived, unnatural modern setting of small holdings with limited grazing access. Keep a health watch going every day on appetite patterns, eliminatory patterns, and general carriage and attitude. The goatkeeper has a minimum of 730 opportunities a year to observe the herd; most of these are moments of well-being. Adding two fecal checks yearly adds to the depth of the health watch and gives care a preventive rather than a crisis-symptomatic focus.

Abscesses

Second to internal parasitism, I would list abscesses as the next major intrusion on caprine well-being in the United States today. My barn has been totally abscess-free, so I only have the experiences of friends and bookish advice to pass on. Most of the latter is at times wildly conflicting. Letters from friends on the West Coast and complaints from other acquaintances about having abscesses crop up after boarding goats from other herds and then being unable to get rid of the problem, have brought the situation home to me in a personal and almost heartbreaking way.

The really widespread abscess problem today seems to be a contagious condition known as *caseous lymphadenitis*, caused by the organism *Corynebacterium pseudotuberculosis* or *Corynebacterium ovis*. It is a lymph system disease that spreads easily throughout the body. Often internal abscesses form, as well as externally visible ones. Antibiotic therapy doesn't seem to be effective, but a general vaccination against *Corynebacteria* has helped some people's herds. I have had no personal experience with this problem, but the current journals and reference books, goat club speeches, and exchanges and convention talk are full of it, and it has become one of the targets, along with coccidiosis, of ADGA-sponsored research. Novices shouldn't try to diagnose or lance abscesses. Get veterinary advice and lab analysis if possible.

If abscesses have appeared and you have no access to professional diagnosis, you have to proceed slowly. Record any appearance of abscesses in your herd. If they appear on several animals, you may have *caseous lymphadenitis* in the barn and should try veterinary consultation by telephone, and vaccination. However, if the abscess turns out to be an isolated occurrence, you can administer an over-the-counter antibiotic such as Pen-strep according to the package directions for small animals, for four to seven days. Common abscesses will often respond to antibiotic therapy by decreasing in size or by heading, breaking and healing. An abscess can also be lanced if it is a run-of-the-mill or isolated abscess. However, a difficulty involved with this is that an abscess can be confused with a hematoma, a blood-filled swelling caused by a direct injury or bruising, which should not be lanced. Check before lancing by drawing a small amount of fluid from the abscess with a sterile, disposable needle and syringe. Don't lance the abscess if the fluid is bloody.

The discomfort of the animal and the bodily position of the abscess should determine your decision about lancing. Lancing should always be accompanied by a systemic antibiotic cover. Cleanse the area by rubbing and clipping the hair away from the abscessed area and scrubbing with hot soap and water. Dry and then rub down with alcohol. Make an incision between one and two inches wide down at the *base* of the bump, not at the peak; peak incisions tend to leave pus pockets due to inadequate draining. With small sterile scissors, clip away a strip of the skin to widen out the drainage channel. Cleanse the abscessed area thoroughly with a soap and water solution, allow to dry (preferably out in the sun and air), and then apply a topical antibiotic. Spaulding recommends "squirting it on."[23] I've never lanced a goat abscess but I've lanced them several times on people and dogs. It is a straightforward procedure. Even when dealing with an isolated abscess, use surgical gloves — cheap irregulars are usually available at hardware stores. The material that drains out of an abscess is loaded with pathogens and the gloves will protect you from making contact with unfriendly staph or strep bacteria. I sterilize scissors and needles in a pressure cooker for a few minutes, and use disposable, prepackaged needles for doing an initial probe to check for hematoma. Some people use sterilized razor blades but I prefer stainless steel scissors and probes. Exposure to sunlight and applications of vitamin E and goldenseal powder, overdusted once new tissue has formed, promote good healing.

For persons far from veterinary services or those who object to systemic use of antibiotics, I recommend the root of *Echinacea angustifolia* (dried) or the fresh root of *Echinacea purpurea* for use in a covering therapy when lancing abscesses, or in reducing or cleaning them out without incision. It may even be useful in combating *caseous lymphadenitis,* but there is no body of controlled experimental data to bear this out. The echinacea root possesses definite antibiotic properties and in my personal experience has been found completely effective in cases of blood poisoning, mastitis, lymphatic system inflammation, surface boils and pustular outbreaks, as well as in highly resistant staphylococcal infections of goats and human beings. After a nurse of my acquaintance cleared a stubborn staph infection she had contracted at the hospital and carried on and off for years through various antibiotic therapies by a consistent program of echinacea usage, I entertained what seemed then a bizarre notion of utilizing *Echinacea angustifolia* for milk goats.

I began by using it as a cover whenever wounds or cuts were sustained. It had the great advantage of leaving no residues in the milk. It is slower, cheaper, and less disruptive than the use of broad-spectrum antibiotics. This year it effected a cure of peracute mastitis in a golden Labrador when no veterinary would see the dog one weekend. Its range of activity is wide; it is active against numerous staph and strep strains and against hard-to-identify pathogens that attack the lymph system.

Dosage should be one ounce per 100 pounds of body weight administered one-half ounce per feeding by directly adding dried or fresh chopped root into the animal's grain. If the animal is off feed or you are resting her from concentrates, brew and steep the herb, one ounce per pint of boiled water, overnight in a covered jar, as you would other herbal infusions, and administer the infusion via the animal's drinking water over the day. This herb is not widely employed in European herbal practice and the reader will find no reference to it in the herbal veterinary writings of Juliette de Baïracli-Levy. It is vital to utilize it in a consistent program for from four to seven days, one ounce per day or more, as above. As with any antibiotic type treatment, it is wisest not to discontinue use in forty-eight hours just because symptoms have superficially abated. It should also be noted that it works slowly, needing time to suffuse the lymph system, and is not appropriate for acute or very advanced inflammatory conditions where time may be of the essence in saving an animal. I have had experiences where it has worked dramatically in such situations, but would only advise it if no other aid were available.

Accidents

Accidents, in my opinion, form the category of next most threatening divergences from goat well-being. Among these I include poisonings, cuts, wounds, bruises, and abortions. In all cases of premature birth it is best to call in veterinary help. In this way, the fetus can be checked for any signs of microbial involvement. This cause of abortion or premature birth is possible but rare.[24] Common symptoms and behavior changes in bred does when abortion threatens are: lack of appetite, dazedness, bloody discharge, and keeping back from the herd. Most abortions I have seen or heard of resulted from fighting, stress caused by ingesting toxic plant material such as mountain laurel, or other accidents. In any event, call upon your vet so that the doe can be treated to prevent possible septicemia and to make sure all the placenta is expelled. If you are without professional help, isolate the doe and keep her quiet, and then observe to see whether or not the placenta has come out. If the placenta has been retained, introduce a uterine bolus designed to stimulate its passage and administer a broad-spectrum antibiotic. Don't breed a recently aborted doe the very next time she comes into heat, even if you lose the season. Give her time to heal and have her hormones get back on course. I have had good luck breeding does that had aborted, out of season, by boarding them with a buck for several weeks. This will yield a deep winter kidding but will produce milk within six or seven months after the aborting incident.

Overeating is a form of accident, but the conditions it may cause are treated under overeating and underfeeding problems.

Surface cuts should be disinfected with hydrogen peroxide, iodine tincture, or clean running water if that is all that is available. If the animal is in good health, no further dressing is needed.

Wounds need more attention than surface cuts. If the wound is a deep puncture wound it should be washed out thoroughly to the bottom. Use a syringe full of hot soapy water and follow up with a disinfectant. In unvaccinated stock, an injection of tetanus antitoxin along with veterinary penicillin is wise. Prevention via vaccination with tetanus toxoid at six weeks or over, or the use of a multipurpose vaccine against clostridial infections, is the best approach. A booster after a year will usually yield immunity for years. Immunization is requisite if the goats are being run on land once used by horses. Another important wound-preventive measure is not to run goats in pastures fenced with barbed wire.

Bruises, scratches and tears should be disinfected as above and topically dressed. I treat bruises exclusively with comfrey, in the form of salve, ointment, or a homemade paste of the leaves chewed or pounded to gumminess. Scratches and tears are dressed with vitamin E squirted from a 100-International-Unit capsule and overdusted with powdered goldenseal. I do this once healing and scabbing have just started. Get wounded or cut animals out in the sun whenever possible for exposure to its germicidal and healing qualities.

Poisonings Salivation to the point of foaminess, weakness, nausea, and profuse vomiting signify the effects of a goat's having ingested poisonous material, usually poisonous plant material. There is no agreement among authorities on just which plants are toxic to goats. Most listings definitely say that wilted wild cherry, rhododendron, azalea, mountain laurel, and yew are highly toxic.[25] Agreement stops there. I have witnessed goats eating Saint-John's-wort, mugwort, wild cherry — both wilted and fresh — mullein, and milkweed with no ill effects. All these appear in some lists of toxic plants. I have seen a young kid eat early (no heads) milkweed, go into a convulsion and then return to normal within twenty-four hours with no foaming and no vomiting. I have never seen a goat ingest mountain laurel, even in the tiniest amounts, without becoming violently ill. Feed your animals more hay than usual when outside grazing offers only laurel (early spring and late fall). I advise not trying to eradicate mountain laurel in the pasture by cutting it back or pulling it up. It will resprout vigorously and these sprouts are highly toxic.

Responses to toxicity vary from individual animal to animal, and from what I have seen, the plant's stage of growth or maturity is also involved. However, we possess no really detailed knowledge on this problem and generally goats do *ignore truly harmful plants*. It is only when they are young and foolhardy, or when little or no other grazing is available, that they experiment with dangerous foods. There is little your veterinary can do in cases of plant poisoning. Keep your animal quiet and provide lots of fresh, warm water. Other poisonings can result from moldy hay or grain, but these are usually minor digestive upsets, often revealed by an isolated feces change. The goats will go off feed for twelve to twenty-four hours and heal themselves. Inspect your hay and grain stores and get rid of spoiled stuffs.

Overeating and Underfeeding

Bloat results from excessive amounts of gas in the stomach. Overeating on lush legume forage, or gorging on grain or (in the case of kids) on milk is the usual cause. The animal gets distended on the left or on both sides, and uncomfortable and sluggish. Keep it on its feet. Exercising bloated animals has helped in the minor cases I have encountered. If distension persists and discomfort escalates you need a vet or the assistance of an experienced goatkeeper or dairy farmer. Prevention consists of always feeding hay before turning stock out onto lush pasturage, maintaining a consistent grain-feeding program and keeping grain stores secured from the goats' maraudings.

Indigestion also arises from overeating or eating moldy foodstuffs. Gas pains may cause the animals to stomp and grunt and nose at their sides in an ungoatlike series of gestures. Feed a couple of table-spoonfuls of vegetable oil to the animal. If a diarrhea and a refusal of feed follows this behavior, you may want to give two tablespoonfuls of china clay (kaolin) in water or Kaopectate. The clay acts as an absorbent for gas and toxins, and helps bind up the system.

Enterotoxemia

Often described as overeating disease, enterotoxemia has a reputation as a leading goat killer. Symptoms are foul smelling diarrhea, lack of coordination, little or no appetite, and weakness in animals that were the healthiest in the herd the day before. There is a multiple vaccine that can be used to protect the herd against enterotoxemia and other *clostridium*-caused conditions. Consult your veterinary and vaccinate. If the herd has not been vaccinated and an animal comes down with these symptoms, call your vet for immediate antiserum or antitoxin treatment. Outbreaks are almost always traceable to an increase in concentrate consumptions, or a too-lush feed of rich pasture or milk. *Clostridia* are normally present in the goat's digestive tract, but a normal fibrous diet keeps the activity of these bacteria at an acceptable level. They can only proliferate and get out of bounds in an airless situation. Gorging clogs the goat's system, setting up a puddinglike trap that has little oxygen. In this clogged mass the *clostridia* can leap to an abnormal increase and with that extraordinary increase comes a great surge in the toxins they produce. It is enough to kill a goat suddenly. Bulky, fibrous feeding keeps order in the digestive tract. Enterotoxemia symptoms are easily misdiagnosed. I have heard of instances in which not even the autopsy of a dead animal revealed *clostridia* and so further losses to the herd were experienced.[26] Don't push your animals hard toward high production by suddenly increasing their grain rations and feeding hay with a high legume content. Overly well-fed goats that are rapidly changed to a new, higher level of rich intake are leading candidates for enterotoxemia.

Ketosis

Ketosis is sometimes described as a disease of undereating. It occurs in pregnant animals close to term. I could categorize it with pregnancy conditions, but it also seems to fit in here. It is a nutritional and metabolic disturbance occurring within the month before freshening. Often referred to as a blood sugar deficiency disease, its direct cause is a carbohydrate or protein deficiency. A number of variant situations can be involved: an underfed doe, an overfed and underexercised doe, or a seemingly well-fed doe that is undernourished for the multiple fetuses she is carrying. Stress and changes, such as moving or severe weather or a change in the ration being fed, may bring on ketosis. It occurs when unusual nutritional demands are made upon the doe. Declining appetite is the first sign. Listlessness often occurs. If you have a doe that is heavy with kid and due in a few weeks and who has lost interest in food and gets up one morning very slowly and just stands with her head down off in a corner of the stall, purchase Ketostix or Labstix from a veterinary. These are treated sticks that react when in contact with the urine of an animal that is suffering from ketosis. If the result is positive, get immediate veterinary assistance. Advanced cases do not respond well.

If you are not in touch with a vet, get a solution of four ounces of molasses or glycerin in warm water into your doe twice daily. This has worked for people I know, but it has to start early. See that your doe exercises, leading her around for walks several times a day if that is the only way. If her appetite picks up and the listlessness disappears, maintain the exercise and sugar regimen till term, and gradually increase her nutrient intake by slight increases in the grain. A bran and molasses mix works well. This procedure is delicate because you do not want to overburden a doe with grain and risk delivery complications. If the original symptoms persist, make sure you get veterinary assistance. It may be necessary to have a cesarian section done to save the dam and kids.[27] When ketosis appeared in a neighbor's herd, a local old-timer recommended I gradually introduce a horse feed with a high molasses content into the ration of my in-kid goats because he thought the dairy ration we had been using locally didn't contain enough molasses and that that may have brought the problem on. I did switch that year and no ketosis occurred.

Milk Fever

Milk fever is a condition that occurs right around freshening and is caused by a lack of available calcium. Falling due to paralysis of the rear legs, loss of appetite, muscular trembling, and nervous excitement signal you to call in a vet. The doe will need calcium administered intravenously. Keep her warm, blanketed, and propped up if necessary to keep her from hurting herself. This problem appears in high producers. If it develops in your does, chances are favorable that it will appear in their daughters and reappear in the original does during their peak production years. Prevention consists of providing good mineral balance in the diet and cutting back a little on the high protein ration during the week prior to freshening. Substitute some bran for the regular grain at this point. Even the most carefully nourished, well cared for high producers can come down with milk fever. If this tendency has appeared in your herd and you are far from a vet, keep in touch via telephone and keep a recommended calcium preparation on hand in kidding season. Call for specific dosage and detailed instructions, since intravenous injection must proceed slowly and on target.

Difficult Kiddings

In a normal kidding, a doe's kid presents its head and front feet first. A rear presenting position is also common, with the back legs exiting first. The only problem with this presentation is that the kid begins to breathe as soon as the umbilical cord is broken. If its head is still in the birth canal, there is danger of suffocation. To prevent this happening, you can lend a light assist once the kid is out up to the shoulders by pulling gently downwards on the legs and helping the doe get the head out. You only need intervene if the kid is stuck and not moving out quickly. Cleanse off the mucus and fluids. If breathing is not normal, while holding the kid upside down by the hind legs, force the remaining mucus out by slapping its side, and dip the cord and slather the navel thoroughly with an iodine solution.

In general, if a doe has been laboring for several hours and straining for more than a half-hour, it is wisest to call for a professional assist or to ring up your nearest most experienced shepherd or goat-keeper. If isolation and other circumstances force you to assist your doe yourself, first try constructing a steep artificial slope in the stall. Place the doe with her hind legs up-slope and her front end down for twenty minutes. She may be able to reposition the kid(s) herself. If not, when forty minutes or so of hard straining have gone by, scrub up to your elbows in soap and water, and cut short the nails of your more skillful hand. Get a friend to help restrain the doe if needed. Lubricate your hand and arm with Vaseline or a similar compound — surgical gloves confuse the messages your fingertips send to the brain. To avoid postparturition uterine inflammation you will have to depend upon antibiotics administered systemically and on disinfecting uterine boluses.

Your fingertips may run into the water bag in the birth canal. Break it and move it out of the way. If the passage seems closed or obstructed even in the relaxed pauses between straining or contractions, dilation may not have started, or it may be ending. Ten or fifteen minutes of examination and repositioning of the head or legs should result in the doe's birthing the kid. If it does not, there may be a further problem that your assistance can't help. Fat or elderly does may be beyond your help. If you can get your hand and arm into the passage, you can aid the doe and the kid.

If labor has been going on for an hour with severe straining for much of that time, and only the front legs are presented without the head visible, the head is probably turned back. Resist the temptation to just pull on the legs. Scrub up, lubricate, and reach in with your fingers and hand to check what has happened inside. Once you have ascertained the position of the turned head, get a grip on the lower jaw and pull the head into presentation position between the legs. Gentle but firm downward pulling in accord with the doe's straining will help the kid out.

Sometimes one leg is folded back. Follow the cleansing and lubrication instructions and place your hand in the canal, tracing from the presented leg inwards with your fingertips. You want to make sure that you are handling the legs of one kid only. You may have to push the head back a little to reposition the legs. Then move it forward and pull gently downwards on the legs. Try to work with your doe's contractions or other motions.

It happens at times that a kid may present itself butt end first with both legs turned back, a true breech position. Scrub well, lubricate, enter the birth canal and work your hand carefully in till you locate the hocks. Push gently forward above the hock and slip the leg back over the pelvic rim, covering the hoof with your hand so as to avoid tearing the uterine lining. Bring the second leg up to face the canal opening. When both legs are facing outwards, grasp and pull down gently.[28]

Unusual presentations can be avoided almost all the time by

proper management practices: good nutrition, adequate exercise, and exposure to fresh air and sunlight throughout the gestation period. Two tablespoons of chopped red raspberry leaves and a tablespoon of powdered kelp should be fed daily in the last sixty days prior to kidding to any does that have had a history of difficult kiddings, and to first fresheners.

Prolapses (eversion)

Vaginal and uterine prolapses do occur in goats, though rarely. They are possibly the result of a genetic factor. References to these conditions are absent in the literature of the natural-rearing school. The primary symptom is a pink, cylindrical protrusion of a small mass from the vulva of an in-kid doe. Caught early, this condition can be treated by confining the doe in a box or narrow stall that slopes her hind end up and her fore end down for part of every day until term. If the protrusion is more than a few inches long, you may need veterinary assistance to place pins in the vagina to keep the tissues in place. Prolapse or eversion of the entire uterus is possible at kidding time, so stay in touch with your vet and have experienced assistance prearranged and available for that week. It may be necessary to cull a doe that suffers this condition.

Uterine Inflammations: Pyometra, Metritis

If your doe goes off feed and you have seen occasional pus discharges at the vulva, take her temperature. If she is running a fever, call your vet. If you are on your own, administer an antibiotic from your emergency aid cabinet. Metritis can result from an assisted kidding where you have had to penetrate the doe's reproductive tract, or from a retained placenta. If the doe does not respond to a program of antibiotic therapy, you need professional help and a lab culture of the discharge so that the specific pathogens can be identified and treated appropriately with a more specialized medication.

Heat Cycle Oddities

If your doe seems frequently in heat at short intervals, there is a possibility of cystic ovaries. A doe not returning to heat once the season sets in may have a possible retention of the *corpus luteum,* a yellow glandular mass in the ovary formed by an ovarian follicle that has matured and discharged its ovum. Veterinary consultation and hormonal injections may resolve such difficulties.

Pink Milk

A heavy producer may show blood in her milk, or "pink milk," in the early weeks of her lactation but have no other mastitis indications. This condition can result from rough milking or just too much pressure in the bag which causes the rupture of small capillaries. Milk her carefully, three or more times a day if necessary, to relieve udder pressure. Massage the bag lightly and use warm compresses twice a day until the milk clears. As a precaution, I feed this milk to the cats or chickens.

Congested or Edematous Udders

You frequently encounter these two udder conditions at the freshening of heavy producers. I treat for the puffy, fluid-retaining, edematous type of bag, and for the congested, overly hard bag by applying warm herbal compresses three times daily, leaving the kids on to suckle if possible (sometimes the udder is too hard and engorged), and by milking three times a day until the conditions ease up, usually within a few days. If there is no improvement, start a close watch of the doe and take her temperature twice daily. Some congestions and edema develop into, or are a sign of, mastitis; some ease off after the stresses of kidding and incipient production lessen.

Mastitis

Any inflammation of the mammary gland, the udder, is known as mastitis. There are infectious, bacterially caused forms of mastitis and noninfectious types that result from injuries. In the latter, soreness and inflammation can be dealt with by hot compressing and massage to soften the tissue, and by frequent milking as outlined above. Goldenseal root and elder flower make good brews for herbal compressing.

Infectious types are treated by veterinary care, usually with an antibiotic regimen. When going antibiotic, I recommend for goats a systemic approach rather than udder infusion. You will probably encounter split veterinary opinion on this question. There is a great deal of controversy surrounding the entire problem of mastitis and you will have to choose which direction you prefer. In my area there is a land-grant college mastitis lab where herd owners can send milk samples for cultural examination. Once particular bacteria are identified, treatment can be quite specific and focused, and this service provides an excellent preventive tool. In addition to all the health measures discussed throughout this book, I would like to reemphasize the role teat-dipping plays in mastitis prevention. An active and sizeable bacterial population is always present in the teat canals. Dipping the teats, or spraying them with a converted window-washing spray bottle full of an effective teat-dipping preparation after milking is completed keeps the basic population within bounds and prevents unwanted transfer. For the small keeper, prevention and a reasonable acquaintance with the basic visible symptoms are usually all that is necessary.

Mild or chronic forms of bacterially caused mastitis are characterized by lumps, bumps, or hardness in the udder. There will be some milk abnormality present, such as clots, blood, or off flavor.

Acute mastitis is characterized by loss of appetite, a hard, swollen udder, abnormal milk, and a decrease in milk yield. Very acute or peracute mastitis will have as signs high fever, weakness and no appetite, very hot, hard and swollen udder, and very abnormal milk containing pus, clots, and blood. It can advance rapidly to a gangrenous condition in the bag if not checked. When these extreme symptoms appear, get your doe and her milk examined by a vet. These conditions are treated antibiotically under close veterinary supervision. Isolate the doe from the rest of the herd, and milk her last so that you do not pass the bacteria to the other milkers. Does on antibiotics must have their milk withheld from human consumption. Observe the milk withholding times of the medications used. Always follow through for the recommended time when undertaking a course of antibiotic treatment, or risk a flare-up. Mastitis is infrequent in small herds. Antibiotic infusion of the udder during the sixty days the doe is dry is usually totally unnecessary. Good attention to disease resistance factors, diet, sunlight and fresh air, combined with scrupulous sanitation and teat-dipping make occurrences rare.

Herbalists argue convincingly against use of antibiotics for mastitis in goats and the mainland Chinese veterinary practitioners make extensive use of herbal preparations for treating mastitis. Information on the Chinese practices is scanty now and we know little about either the specific plants used or their methods. Baïracli-Levy cites methods involving imposing grain fasts on animals, and using compresses and massage. I would intensify the compressing if the doe is running a fever and also make use of the echinacea preparations I have described in the section on abscesses.

Urinary Calculi

A difficult problem that besets male goats is the formation or deposition of stones, commonly calcium crystals, in the narrow urethral tract. The stones often lodge at the tip of the penis and can be removed by veterinary surgery. Symptoms include a strained, hunched stance when the buck attempts to urinate, declining appetite, listlessness, and loss of sexual interest. Debate is extensive about the causes for urinary calculi. So far, I have encountered the following hypothetical causes: barley feeding, hard water, too high a concentrate regimen, hereditary mineral imbalances, and metabolic imbalances. Herbalists recommend utilizing wormwood or hydrangea root to dissolve calculi. Other recommendations include acidifying the buck's drinking water with cider vinegar or cranberry concentrate in an attempt to acidify the bladder and urethral tract. If removal of tip stones is done, have them analyzed. There are myriad kinds of stones and an analysis of the mineral content may clue you into the constitutional metabolic imbalance you are encountering.

If removal of these stones or herbal dissolution do not work, your buck may have a congenital problem that will promote more stone development, often further up the urethral tract. Some owners ask their vets to attempt further surgery on the animal, but this may impair or end breeding capability. A buck that responds to none of the treatments and has to be repeatedly catherterized to eliminate urine will eventually fall to uremic poisoning. He becomes autotoxic by absorbing his own wastes since he can't eliminate normally. Diagnosis and consultation with a veterinary surgeon are imperative, but you should be aware that an animal that goes on developing urinary tract calculi cannot be half your herd and will probably be better off humanely put under. Having experienced this particular problem through the agony of an eleven-month-old purebred, I can now say that he should have been spared the last month's duress. We were ever hopeful because the condition is so rare in young animals, and this particular buck never even ran a fever or lost his appetite right up to the uremia stage. But autopsy revealed he was the victim of some hereditary difficulty, for he had many calculi all the way up the urinary tract.

Rots, Ticks and Other External Parasites

Talking of bucks brings me around to a final class of problems. The only other health problem I have encountered in bucks is their frequent endurance of lice, foot rot and other external parasites. It seems to me that they are more often the victims of these plagues and vermin than does for several reasons. They are not observed and tended to as assiduously as the does, they are often kept in close confinement for long time periods and they can grow to proportions and reach temperamental states where they can be difficult to handle and groom.

Fungal Inroads Foot rot is a fungal infection that can pentrate goat hooves. Wet bedding, wet ground, and close confinement provide its means. Keep feet trimmed and inspected monthly and your goats exercised, even through bad weather. If foot rot develops, you may see lameness or swelling. Soak the feet in a hot Epsom salts solution and trim the hooves down to get rid of the putrid matter. Pack with a paste of vitamin E and powdered goldenseal.

Ringworm is another fungal condition that can show up in animals that have had a long, enclosed winter. Advanced cases look horrible, with crusty circles showing around the body, face, and neck. Bathe afflicted places with a goldenseal infusion (one-half ounce of cut root steeped overnight in a pint of boiling water), cleaning off crusty matter and saturating areas with the goldenseal brew. Then dress with vitamin E overdusted with powdered goldenseal. Keep animals in isolated stalls, but get them out in the sun whenever weather permits. Sunlight is an excellent fungicidal agent.

Lice Dry and patchy-looking skin, dandruff, and constant rubbings and scratchings, itchings and twitchings, combined with hair losses are indications of lice infestations — again a confinement plague. Exposure to sunlight and fresh air usually controls or eliminates infestations. If the itching and scratching continue and the weather does not allow for much outside romping, secure a louse powder made for dairy animals, or use plain powdered rotenone, and dust the whole herd weekly for several weeks. Make sure the powders are safe for milk animals, and follow withholding directions if there are any. I withhold the milk from human consumption for three days when utilizing rotenone. Quassia chips around the stalls also help keep the critters down.

Mites Mites cause mange which is signaled by bald patches and pimplelike outbreaks followed by crusty areas, and accompanied by much itching. It can be a serious and highly contagious condition and it can look like many other things. Have a veterinarian do skin scrapings and then follow his or her advice for treating the animals. I recommend that you doublecheck on the ingredients of any preparation you set out to use for mange. In particular, watch out for lindane content. Lindane is a chlorinated hydrocarbon, a DDT relative, dilute solutions of which are much used in combating mite infestations in

126

humans and animals. Do not use it on your goats, since it is taken up by the body, stored in the fat tissues, and excreted in the milk for a long time. Close clippings, consistently repeated bathing of irritated areas with goldenseal infusion, and the use of a sulfa ointment such as the ones prepared for human use in combating scabies (mite infestation), have gotten rid of a mange outbreak that was caught early in a friend's herd. And once again, get the herd into the sun.

Ticks Ticks are big, easy to spot, external parasites. A Texas-born friend recommends never pulling off a tick. Remove them by coating the tick with nail polish (it suffocates) or by burning the tick with a lit incense stick — they drop away to avoid immolation. These are tried and true tick-country remedies. I have also seen people pull ticks off slowly with a twisty, rotating kind of motion so that the head and whole pest come away. If you keep goats in tick-infested country, keep short-haired, short-clipped goats. Severely tick-infested goats can be treated with a rotenone solution.

After such a lengthy list of possible troubles, you may be wondering about all that well-being I said was the rule way back at the beginning of this chronicle of possible woe. What will the goatkeeper see in the small herd? External parasites are occasional. You can expect to encounter yearly upsurges of internal parasites in most herds. There seems to be less coccidiosis in climates that experience severe winters, but that is not an absolute. Kidding difficulties are infrequent, as are ketosis and milk fever. In my experience, edematous udder and congested udders occur often, particularly among heavy producers and their first-freshening daughters. Mastitis has occurred twice in isolated instances in my small herd, but it is more frequent in larger herds that are kept inside and milked by machine. We have to expect that genetically improved high producers will have a greater share of these types of problems; it just is not natural for a goat to produce sixteen or eighteen pounds of milk a day.

We have never had a coccidiosis, tetanus, enterotoxemia, or *caseous lymphadenitis* outbreak. We keep the number of our goats low and have them stalled separately and eating out of individual pans and hay racks, and we allow our goats to do no hobnobbing with does coming in for buck service or with other goats at shows or fairs. From a health and contact point of view, they live in a sheltered environment. Through the occasional visits of does coming to be bred, we have had some minor problems brought in — such as goat pox[29] — but nothing major. We have seen our share of accidents including two abortions brought on by fighting, some plant-poisoning incidents, and minor cuts and wounds. Pneumonia, which obstructs breathing, has never occurred, even though the goat books are unanimous about the ease with which goats can be stricken by it through stress, parasitic inroads and microbial activity.[30] The goats here endure a climate that can have sixty Fahrenheit degree temperature differentials occur within a twenty-four-hour period, that is legendarily damp (many humans hereabout suffer from Catskill catarrh), snowy, and wildly variable. Thus far, they have met all the challenges, and are as content in 15°F.(-26.1°C.) weather as they are on an 85°F.(29.4°C.) summer day. With a well-being record and vitality such as this, dairy goats turn out to be a near-perfect starter animal for newcomers to small livestock.

CHAPTER 11

THE GOAT
IN THE
KITCHEN

It is proverbially said that bringing a goat into your house dispels all your troubles. Perhaps the folk wisdom behind this is in the fact that a goat within your dwelling space will keep you so busy and distracted that you will not have time to think about other woes. I have met up with goats in other people's kitchens, usually young kids enjoying a bit of household warmth and attention during the weeks of deep winter weather common in my region. However, the more ordinary way for the goat to come into the kitchen is by way of her milk.

Even today, recipes that call for refined white flour, sugars, and even canned soups as casserole and sauce bases abound in dairy goat literature. Homesteaders go to a lot of trouble to provide their tables with fresh, raw, unadulterated milk, cheese, and yogurt. It seems totally contradictory to add these nutritionally superior, hard won, and delightful-tasting products to canned soups in order to whip out a meal! The paucity of goat's milk recipes that have a whole foods orientation has caused me to do wide-scale experimentation in my own kitchen. The results of this activity are the basis for this chapter.

There is little attention given to meat because we eat little meat. Instead, cheese will be given much attention.

When I first became a goatkeeper almost six years ago, there was little information available on home-scale cheesemaking. Over the last three years there has been much exchange on the questions of cheesemaking and the daily use of goat milk in the many publications concerned with food quality and small-household production. However, there are further yogurt and cheese pointers incorporated into this chapter because cooking with goat milk is slightly different from cooking with cow milk, and my handling is somewhat different from that of other printed material now appearing.

Butter is only briefly covered here because the goatkeeper needs a cream separator to make butter easily and efficiently and a sizable milking herd to consistently produce a surplus of cream. This is not usually the small-scale producer's situation. The buttermaking material I have included is based on an approach that requires no separators, churns, or other special equipment.

As soon as you bring the milk down from the barn, chill it. The container can be dipped in iced water or you can run cold water over it, or stand it in a shady spot in a rapidly running stream. The objective is to reduce the temperature of the milk from its initial 90° to 100°F. (32.2° to 37.8°C.) to the 40°F. (4.4°C.) range. This reduces the chances of bacterial growth and subsequent spoilage. Raw milk keeps well under refrigeration for about seventy-two hours. After that, bacteria will begin to proliferate even under refrigeration, and make the milk taste "goaty." If you expect to have milk standing for more than three days, plan on freezing or culturing it. Goat's milk freezes very successfully and does not separate upon thawing. Thaw it slowly when you need it, and blend or shake it well before using.

Yogurt

Culturing milk has been a traditional method of preserving milk in many areas of the world. In addition to having storage and keeping advantages, cultured milk is more easily digested by the human body than is plain milk. Many culturing agents are bacteria that are highly beneficial to the human digestive tract. Numerous yogurt-making recipes and how-to-do-it descriptions can be found in the natural foods literature. Since I use about one gallon of yogurt every two to four days, I have developed some simple, workable, energy-efficient approaches to yogurt making.

My easiest approach to goat yogurt production is to heat four quarts of goat milk to below scalding temperature and then to allow the milk to cool to lukewarm. I then place six tablespoons of a high acidophilus yogurt (available in most health food stores) into a glass gallon jar and slowly pour the warm milk into the glass jar, mixing the starter culture throughout. I then cap the jar, wrap it completely in a towel and place it on a warming shelf, or place the unwrapped jar in a Styrofoam picnic hamper and weight down the lid. If left to its own devices, the yogurt will culture overnight or sooner, depending on the general room or atmosphere temperature. I refrigerate without disturbing or mixing the contents. This method sets up a fine yogurt, without the use of dried milk powder, agar-agar, or gelatin. Once mixed or dipped into, the yogurt may start to get runny. America has gotten conditioned to yogurt that is thick, custardy, dessertlike, or frozen. Few people are aware of the stabilizers and additives being used to achieve these consistencies. Yogurt from the Middle East or Eastern Europe is not solid but runny and loose textured. In our household we have reaccustomed ourselves to the thinner, fine-textured, cultured product made without additives.

Another approach to yogurt relies upon a specific culture that cultures raw milk at room temperature with no preheating required. The PIMA culture is a freeze-dried, Finnish-derived culturing product produced by persons associated with the Price-Pottenger Foundation. The great advantage of PIMA is that it cultures milk quickly and directly with no fuss or elaborate preparations. There is not even an absolute need for a stove or refrigerator, or for any fuel consumption. We have employed this culture to produce yogurt for two to three hundred persons a day over conference weekends. A batch is started by emptying the contents of the packet into one cup of raw milk at room temperature and allowing it to culture for thirty-six hours. Larger and larger batches may be made by adding two cups of the resulting culture to each gallon of the raw milk you plan to culture. Refrigerated milk or any milk cooler than 65°F. (18.3°C.) must be heated to room temperature. Though the PIMA instructions state that goat milk cannot be used for the initial batch of culture, I have used it and have produced both a successful starter batch and many successive gallons of yogurt.

In the winter, yogurt will need a special warm spot in your heart and home. The culturing bacteria will not be active if the nighttime temperatures drop to 55°F. (12.8°C.) or colder. Your yogurt making will need a warm shelf near a stove, a place atop a water heater, or a hot-water bath of its own. Inevitably, culturing will take longer in the winter and will move rapidly in the summer when temperatures in the house reach 80° to 90°F. (26.7° to 32.2°C.). One has to become sensitive to these seasonal changes and avoid thinking that yogurt always takes the same length of time to set up. It is a process that involves various living organisms.

Goat milk yogurt is excellent as a salad dressing or mixed with a compote of dried fruit. It serves as a fine dressing for baked potatoes and improves the taste and texture of mashed potatoes. Potatoes and milk products complement each other's amino acid contributions and thus provide vegetarians with a complete protein when used together.

During the summer, yogurt blended with fresh fruit or with honey and wheat germ produces a homemade version of a convenience meal in a glass — the East Indian "lassie." Lassies are best when made from very ripe soft fruits such as bananas, berries, and peaches. A dash of cinnamon, clove, or mace brings out the yogurt tartness and the fruit sweetness. East Indian yogurt soup is also quickly made up by blending half a cucumber, three tablespoons of lemon juice, and a pint of yogurt. This is usually refrigerated and served cold. It also makes an excellent, soothing accompaniment to hot curried dishes, summer or winter.

Winter uses for yogurt include its addition to sauces and curries. It can be a major baking ingredient, as in the following cornbread:

Cornbread

1 cup yellow cornmeal	1 cup goat yogurt
1 cup whole wheat flour	2 teaspoons honey
(bread or hard wheat)	(optional)
½ cup raw wheat germ	3½ ounces maple syrup
1 teaspoon salt	(just under ½ cup)
2 teaspoons baking soda	2 tablespoons oil
2 beaten eggs	

Combine the dry ingredients and mix well. Mix the wet ingredients in a separate bowl and add them to the dry ingredients. Mix well. Pour into a shallow, well-greased pan and bake in a 350°F. (176.7°C.) oven for 20 to 30 minutes or until the top is golden brown and a fork comes out of the middle with no particles clinging to the tines.

I always substitute one cup of goat yogurt for part of the liquid called for by standard whole wheat bread recipes. It improves the texture of baked foods immeasurably. A fine breakfast complement to hot whole grain cereal can be made by soaking several types of dried fruit in water overnight and then puréeing them in a blender. Fill a small cup with alternating layers of the fruit compote and yogurt for a satisfying and protein-expanded breakfast accompaniment. The energy, vitamins, and minerals supplied by the fruit will be enhanced by the vitamin, mineral, protein, and fat contributions of the yogurt.

Goat milk cookery differs slightly from cow milk cooking because of the differing behavior of the fat particles. Goat yogurt and cheese take slightly longer to set up and custards need another egg to firm them up or require more baking time. Cakes and breads develop very fine texture and may need an additional five minutes baking time. All these usually rich foods possess the bonus of being more easily digested than the same foods made with cow milk. Many natural foods theorists agree that goat milk is not as mucus-forming as cow milk, even violently antidairy purists such as macrobiotic proponents. The whey is highly recommended for persons suffering stomach acidity. It is used either straightforwardly as a drink or more indirectly as an ingredient in soups and stews. Substituting part or all whey for cooking water when preparing whole grains adds both taste and nutritional value.

Goat's milk whey forms the basis for an unusual cheese called "gjetost." It is easy to prepare if you have a wood-burning stove. After separating curds and whey in cheesemaking (see below), place three to four quarts of the whey in a heavy stainless steel or cast-iron pot, and allow it to simmer until it evaporates to a pint of dark liquid. Add one cup of whole milk and continue simmering until the liquid reduces to a pint again. By then it has become thick and pudding-like. Remove the kettle from the stove before granulation or foaming occur. Gjetost has a surprising sweet-salty taste. Some people adore it at first bite and others immediately detest it and never change their minds. It makes an excellent spread for whole grain breads and dark rye crackers.

The goat contributes to protein-rich main dishes, even in households that don't eat goat meat. One of my early discoveries was cauliflower-goat pie:

Cauliflower-Goat Pie

1 head cauliflower	2 cups goat milk
3 tablespoons butter or cooking oil	1 cup grated, hard goat cheese
3 tablespoons whole wheat flour	4 eggs, separated nutmeg, salt to taste

Steam the cauliflower until tender. Mash and drain in a colander. Make a cream sauce by melting the butter in a saucepan over medium heat. Add the flour and stir with a whip until well blended. Stir in the milk. When the sauce thickens, add the cheese. Remove the pan from the flame and add the beaten yolks and cauliflower. Allow to cool and add a dash of nutmeg.

Beat the egg whites until they form soft peaks and fold into the cauliflower-cheese mixture. Turn the whole into a lightly greased 9-inch, deep dish pie pan and bake at 350°F. (176.7°C.) until the top is golden and a knife inserted in the center comes out clean. Serves 6 gluttonously and 8 reasonably.

It is delicious served hot or chilled and forms the core of a substantial meal. The cauliflower-goat pie makes a fine meal accompanied by a large green salad and whole wheat yogurt "nan."

Nan is a type of quick, East Indian leavened bread:

Whole Wheat Yogurt Nan

1 tablespoon yeast	*½ teaspoon crushed*
2/3 cup warm water	*cardamom seed*
1 cup yogurt	*2 cups whole wheat bread*
2 tablespoons olive oil	*flour (more may be*
¼ teaspoon salt	*needed)*

Prove out yeast by adding it to the warm water and allow to stand until foaming and rising occur. Combine yogurt, oil, and salt, and add to yeast mixture. Add crushed cardamom seed and flour until a workable dough that does not stick to the sides of the bowl is formed. Knead gently on a floured board for several minutes, then place in a lightly greased bowl. Cover and allow to rise for 30 to 60 minutes (depends upon how much time the cook has). Cut the dough into 8 equal pieces, pat into rounded and slightly flattened shapes. Bake at 375°F. (190.6°C.) for 20 minutes.

Another fine main dish that comes to table frequently at our house is an adaptation of an old French country dish, *"gratin dauphinois."* It combines milk and potatoes, and in the original, some shreds of pork, either as smoked ham or sausage. A meatless, goat-based version is made as follows:

Meatless Gratin Dauphinois

3 pounds parboiled
   potatoes, sliced and
   drained
semicured goat cheese

tarragon or other herbs to
   taste
2 cups goat milk

Layer the potatoes in a buttered baking dish with thin slices of cheese, sprinkling tarragon and a bit of grated, aged goat cheese on every other layer. Pour the milk over all, and top with a thin layer of grated fresh goat cheese. Bake, covered, for 20 minutes at 350°F. (176.7°C.). Remove the cover and bake for 10 more minutes until the cheese on top melts and turns golden.

Pizza can serve as a well-balanced, protein-rich main dish when prepared at home without commercially processed ingredients.

Goat Pizza

1 cup goat whey warmed
   to 100°F. (37.7°C.)
2 tablespoons olive oil
¼ teaspoon honey
1 tablespoon dried yeast
⅓ cup warm water
2 cups whole wheat bread
   flour
1 quart Italian tomato
   sauce (I used home-
   canned sauce)

toppings: onion and green
   pepper rings; sliced
   mushrooms; bean paste;
   sliced olives; herbs:
   garlic, oregano, basil
1 pound soft goat cheese,
   grated

Add olive oil, a pinch of salt, and honey to the whey. Mix and let stand. Dissolve yeast in water and allow to foam and rise. Add to whey mixture. Slowly add 2 cups of flour to the moist ingredients. Our household prefers the breadier crust that develops from using hard red winter wheat flour; soft wheat flour may also be used. Add enough flour to form a dough that is elastic and yet comes away from the side of the bowl easily. Flour a board and then knead the dough for 3 to 5 minutes. Place dough in lightly greased bowl, cover with damp cloth and allow to rise in a warm place for 45 minutes.

After the dough has risen, turn it out onto a lightly floured board and knead lightly for about a minute. Lightly grease 2 cookie sheets or baking pans. Divide dough into 2 equal pieces and roll each out using a rolling pin with a very light touch. Carefully lift dough and place in the pans, gently pushing the dough toward the edges of the pans.

Heat the tomato sauce. Spread whatever vegetables or herbs you're using on the dough, lightly ladle on about 1 cup of sauce per 10-inch pie, and sprinkle the cheese over all. Bake pies in a preheated 400°F. (204.4°C.) oven for about 20 minutes or until the cheese melts.

This type of pie harks back to thick-crusted, country-style Sicilian pizza and is totally unlike the white-flour, wafer-crusted pizza found in America. By adding a thin layer of cooked and puréed beans (pintos, kidneys, or adukis) before adding the sauce and cheese, a very satisfying, protein-balanced main dish is created. Although I add garlic, onions, basil, and oregano to my tomato sauces before canning, I have found that adding a sprinkling of these herbs to the pie before baking is a flavor enhancement.

Butter

Two fundamental cooking ingredients in our household are butter and cheese. Buttermaking becomes more important as herd numbers increase. Small amounts of butter can be produced from cream separated without the need for special equipment, but our two does will not provide enough cream year-round to supply all household butter needs. Once a household begins keeping four to ten milking does, buttermaking can be undertaken consistently, but a cream separator is a must. If you are lucky, you may unearth an intact centrifugal cream separator and not have to pay dearly for it, but they are becoming increasingly rare and expensive on the secondhand market and prohibitively expensive new. A capacious churn will also be needed. For occasional, small-scale buttermaking you can follow these simple steps:

Clean and rinse thoroughly a very wide but shallow bowl, or several if you have much milk. Pour strained, but still warm (not chilled) fresh milk into the bowl(s). Cover lightly with brown paper or freezer wrap, and place bowl on refrigerator shelf. After approximately one to two days the cream will rise to the top. You can skim it with a skimming spoon, a shallow spoon with holes that allow the skim milk to drain. You can make a homemade version of such a skimmer by drilling holes with a jeweler's drill into a bamboo rice paddle or other such shallow, flattish and wide wooden cooking tool. Most of the cream will stick and stay on the skimmer, but the skimmed milk still has some cream content.

Roughly speaking, using a centrifugal separator will yield a pint of cream per gallon of milk. Such figures are vague because the yield depends upon each doe's individual butterfat capability, her diet, the efficiency of the separating technique used, and the stage of the doe's lactation. It is this variability and the difficulty in achieving a high cream/milk ratio that makes butter production for the goatkeeping homestead not too viable. Usually there is not enough surplus milk, enough time, or efficient enough equipment to yield a constant supply.

However, small amounts of butter can be made without elaborate investment or special equipment and the taste treat is worth the effort. Once the cream has been separated and skimmed off, it immediately has to be churned, with a regular churn, or by use of make-do methods — handbeating with a whisk, use of a low-speed electric blender, or by using an electric mixer set at a low speed. In the summer, churn your cream at temperatures between 52° and 60°F. (11.1° and 15.6°C.), and in cooler seasons at between 58° and 66°F. (14.4° and 18.9°C.). The length of time required for churning and the eventual texture of the butter depends on bringing the cream to these temperature ranges. You can adjust the temperatures by placing the cream container in a water bath of the appropriate temperature. Beat, whisk, mix, paddle, or churn the cream in a regular, rhythmic way. If you are using electrical appliances this is not possible, but keep to the lower speeds. A sudden textural change in the cream will alert you to the progress of the cream. Add some cool water, if the season is warm, or some 60°F. (15.6°C.) water if the room is cold. Continued mixing will result in a granular, beady texture. Stop and drain off the buttermilk (save and use), then add cool water, beat gently some more, and drain again, till the water runs off clear. When the butter beads are completely washed, a small amount of salt (I use only one-fourth teaspoon of sea salt per pound of butter) may be added. When butter is made in very small amounts there is no preservative need for the salt, for the butter will be rapidly consumed. Salting is a matter of taste. Some people color the plaster-white goat butter at this stage with Annatto, but carrot juice also works as a colorant. I do not bother coloring it at all. We know that goats process carotene so completely that their milk, cheese, and yogurt are naturally stark white, so there seems no real reason to color the butter just to make it seem closer to cow butter.

Beat the butter well on a drain board to work out any remaining water. The keeping quality of the butter depends on expelling all the water. Fluted boxwood pats called "scotch hands" are the traditional tool employed at this stage to work the water out. I purchased mine in England years ago, but they are now available at specialty cookware departments and shops in America. Wooden spoons will do, but can crack and splinter under heavy usage. Usually, large quantities of butter are cut up in small units of some convenient size, wrapped, and stored under refrigeration. Small amounts simply disappear and storage is never a problem. Glass half-pint jelly jars make convenient storage containers for in-between amounts of butter. I avoid setting milk in plastic bowls to separate and never store butter in plastic containers.

Cheese

Cheese is a fundamental cooking ingredient and protein contributor in our kitchen. By producing and storing cheese it is possible to have your goat's produce in your kitchen the year-round. Dr. Weston Price found that isolated Alpine villagers lived wholly and healthily on a diet of raw milk products, particularly cheese, and whole grain, mostly homegrown rye. These people had no incidence of tooth decay or palatal deformations, and had unusually sound general health. In areas where food must be stored against inclement winters during which little or no fresh food is produced, cheese offers an excellent solution. Six or seven years ago there was not too much information available for the home cheesemaker. The situation has changed; almost every homestead-oriented publication now carries information about cheesemaking.

Several times weekly, I take out my enameled, twenty-quart canning pot, rinse it, and place it over a stove burner. Three to four gallons of sweet, fresh milk are poured into the pot and the contents are heated until they are brought to 86°F. (30°C.). I use a cheap thermometer to test the milk temperature. If the milk has overheated, I let it cool down to 86°F. (30°C.) on its own, or add some cold water or milk if pressed for time. One rennet tablet dissolved in one-half cup of water is added to the warmed milk.

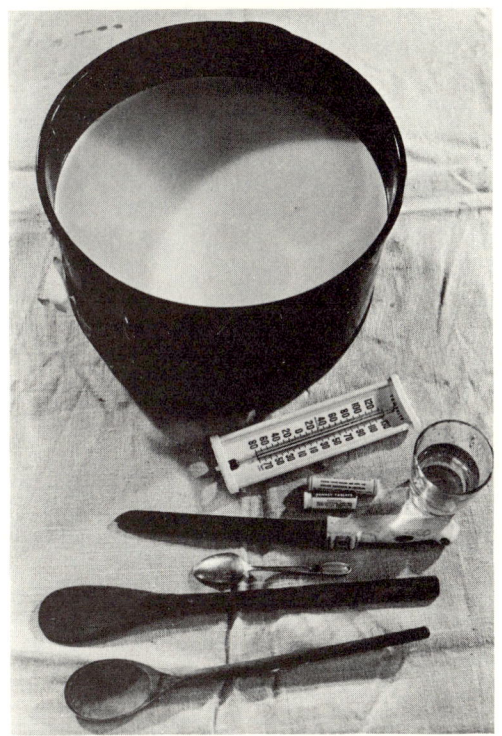

Implements needed for home cheesemaking.

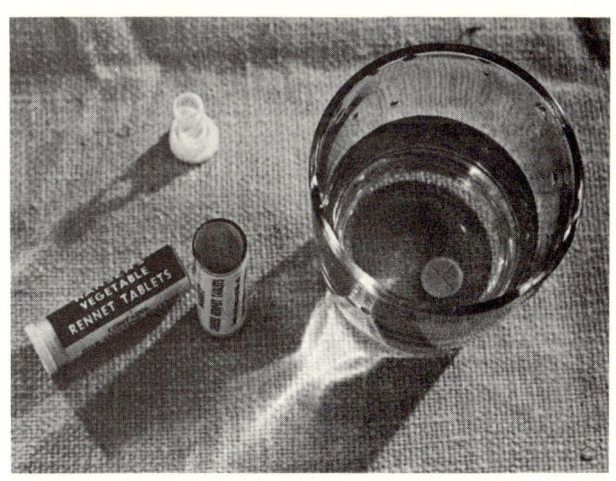

Dissolve rennet tablet in cool water.

Add dissolved rennet when milk is at 86 degrees F.

Most rennet is derived from the lining of a calf's stomach and is a curdling enzyme active in the 80°-to-100°F. (26.7°-to-37.8°C.) temperature range. There are "rennets" derived from nonanimal sources available commercially. All nonanimal curdling products are more expensive than ordinary rennet with the exception of curdling herbs you may gather yourself, and all, in my experience so far, set a less dense curd and required a longer set-up time than animal rennet. They *are* effective and produce fine-tasting cheese with goat milk. (Rennet source information is included at the end of the resources.) I have used three different "vegetable" rennets, each of which had its own idiosyncracies. Most of the time I use regular rennet. The usual recommendation for use is ¼ tablet to each gallon of milk. I use slightly more for curdling goat milk, 1 tablet for 3 to 3½ gallons, and 1¼ tablets for 4 to 5 gallons of milk. Goatkeepers I know who keep Nubians usually make their cheeses with ¼ tablet per gallon of milk.

The curd will usually set up within the space of twenty to thirty minutes, but it can take up to forty-five minutes with goat milk. When the curd is firm and clearly separated from the thin green-white whey liquid, I slice it through and through with a long-bladed knife, cutting the mass into many small, evenly sized cubes. The curds then sink to the bottom of the pan. At this point I wash and dry my hands and then plunge them into the cheesemaking vessel. By hand, I break the curds up more, with a gentle squeezing motion. After the curds sink back again to the bottom of the pot, I set a large

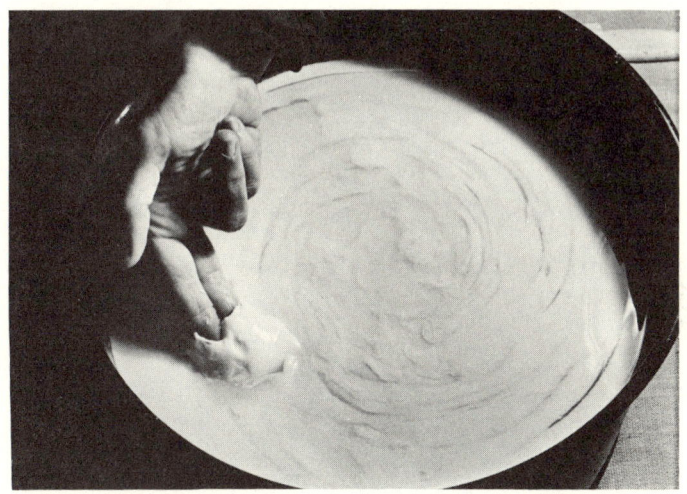

This curd has set and is ready to be cut up.

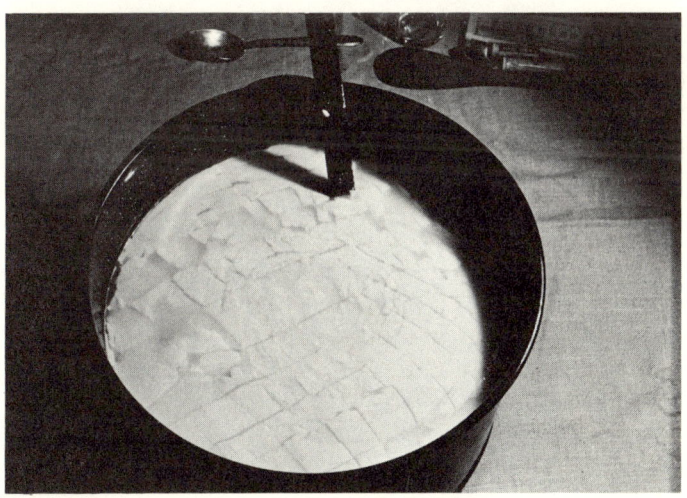

Cut firm curd into small, even pieces, then allow to settle.

141

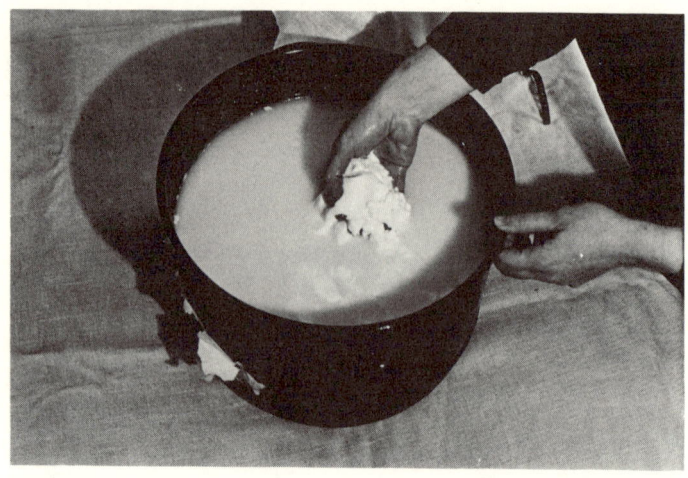

Wash hands, then break up settled curd into even smaller pieces; allow to resettle.

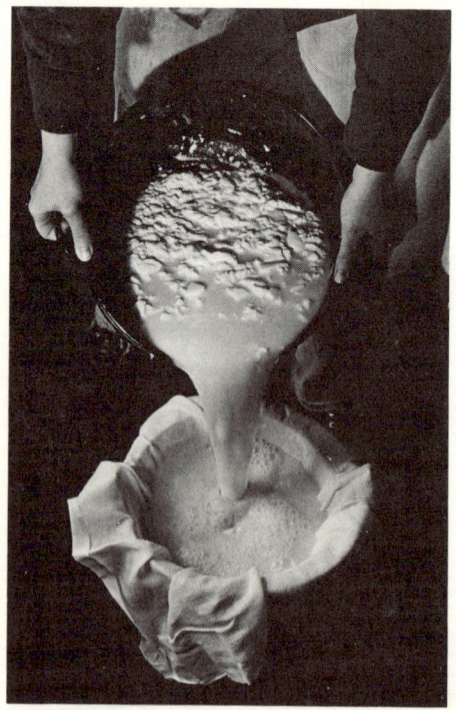

After curds have resettled, pour off whey.

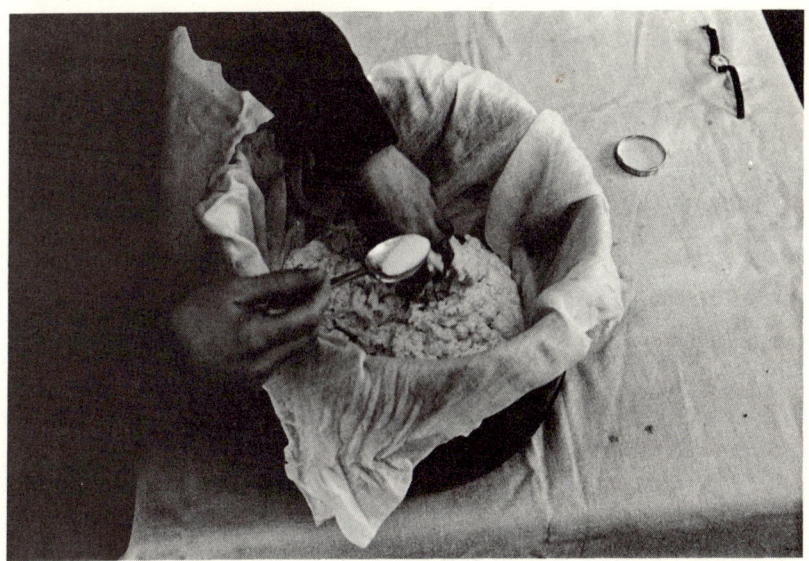

Add ½ tablespoon of sea salt per pound of curd and mix well. Herbs may be added at this stage too.

colander lined with a cotton bird's-eye diaper over a capacious bucket. These diapers are much finer textured and more durable than most of the cheesecloth I have used for cheesemaking. Pampers are pushing hard on the cotton diaper but old-fashioned diapers can still be found with some energetic research. I pick up the pot of curds and whey without disturbing the curds settled on the bottom, and pour the whey into the bucket, catching all the curds in the diaper-lined colander. This is one smooth operation that has to be done without agitating or redistributing the curds. I then pick up the diaper and squeeze out what additional whey I can. I add 1½ tablespoons of sea salt to the curds and, at the same time, any herbs I wish, and then I tie off the diaper and hang the ball of curd in a high and dry place.

I leave the cheese to cure for a week or longer, depending upon the type of cheese I want. When I take the cheese down, I run it under a light stream of warm water till the cloth is wet and pulls away from the body of the cheese without resistance. For a well-aged cheese, blot the cheese and rub it down with a light coating of vegetable oil and allow it to cure further on a rack in a dry spot. Turn every week and check for possible mold. After six weeks the cheese can be placed in the back of the refrigerator and forgotten till next year. For soft,

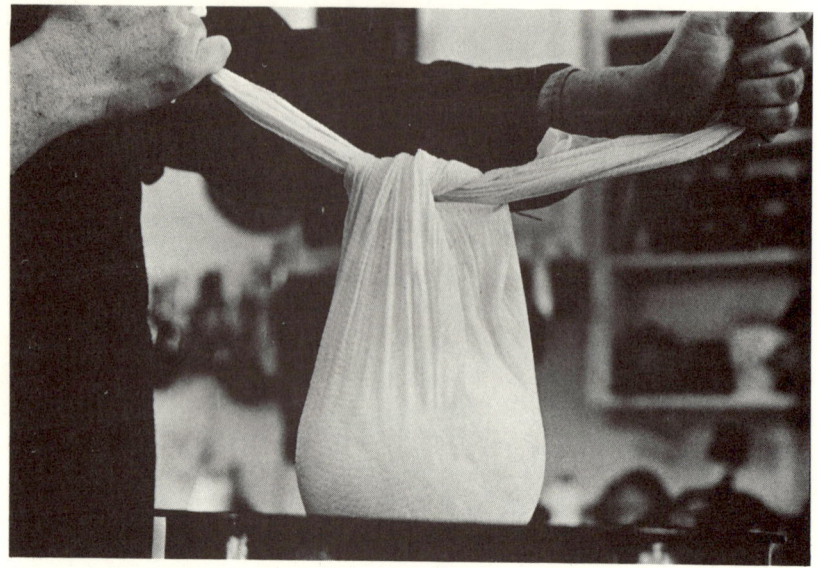

Tie curds off in a cheese bag, cotton diaper or two layers of cheesecloth.

semicured cheeses, dry the cheese gently and do not rub it down with oil. Wrap the cheese in another cloth and store it in a bowl in the refrigerator for use as needed. We use these semicured cheeses for everyday eating and cooking. Many of our everyday favorites are less than a month old. Some of them are even younger, only a few days of age, and come close to cottage cheeses or ricotta in taste but are firm textured. I also salt slices of fresh cheese and store them in brine made of one tablespoon sea salt to one pint of water. This cheese is used in salads and in cooking. Cottage cheese is made by draining the whey off, salting the curds minimally, and then packing into wide-mouthed glass jars. It is stored in the refrigerator for use over the next few days.

The aged cheeses will dry into good-quality, hard grating cheeses within four to five months. They become sharper and more tasty with further aging, but I do not leave them unrefrigerated, for they are subject to much molding in the humid and changeable climate of the mountains where I live. When I set them in the back of the refrigerator for a year's aging, I usually wrap them up in brown paper bags.

Cheesemaking by this process is simplicity itself and requires no fancy equipment and no pressing, fussing, or paraffining. An enormous range of cheese can be produced by this method. I am indebted

Squeeze bag tightly to drain remaining whey.

Hang the cheese up to drain further and to cure.

for the germ of the idea to John Seymour's passing mention of it as an old Scottish highlander's technique at the end of the cheese discussion in *Farming for Self-Sufficiency.*

In addition to producing cheeses that are different in character due to the length of the curing period, I am able to produce variety by introducing various cultures to the curds. If I wish to make a blue cheese or a tilsit, I inoculate the curds before hanging them out by adding shavings of a mother cheese in the same manner as that in which a starter is introduced to a new batch of yogurt. But watch out for blue cheeses! Two years ago I inoculated some curds with fine shavings of a good-quality French Roquefort and then had blue cheeses turn up every succeeding time I made cheese for several months. The spores of the Roquefort mold had escaped and set up housekeeping in the stucco ceiling of my kitchen. New cheeses came out blue as the Roquefort spores found new curds to conquer.

I also utilize a wide range of herbal ingredients to produce unusual and tasty cheese. Raw, minced garlic, one clove per pound of curd, produces a rich and surprisingly mellow cheese that is a leading favorite among family, friends and customers. Be careful to use fresh, firm garlic cloves. Moldy or old garlic may bring unwanted cultures and spores to the new cheese. Sautéed onion makes a delicious cheese addition, as do fresh chopped chives and green onion tops. Two other favorites are sage and caraway cheeses. I add one teaspoon of finely crumbled dry sage leaves to each three-to-four-pound ball of curd; others may prefer larger amounts. One tablespoon of caraway seeds per three-to-four-pound curd ball produces a delicate cheese. Small amounts of sage, thyme and garlic will combine well to produce a cheese slightly reminiscent of the best cured sausages.

Two or three tablespoons of yogurt or buttermilk added to the drained curds produce tangy, fresh-tasting cheeses that combine particularly well with scallions and chives. I also make a good cream cheese spread by hanging a gallon of goat's milk yogurt out overnight in a diaper bag to drain. The fresh yogurt-cheese left in the cloth in the morning is delicious and very creamy.

Before leaving the cheesery, I would like to discuss the apex of goat milk cheesemaking: producing the classic French-style *chèvre,* a fine-textured, dry, unpressed cheese produced in small quantities by traditional methods. I am indebted to Mme. Danielle Haase-Dubosc Gloag for her first-hand research and her observation of Mmes. Roux and Litaudon, French countrywomen, working at their craft in the locale of La Roquelle, Burgundy, France. Modern pressures seem to be reducing the numbers of people engaged in the old processes in that area. The renewed interest in goatkeeping and home cheese production occurring in America may serve to help keep alive and disseminate these old French methods — and a truly great cheese.

Take 2½ gallons of sweet fresh goat milk and heat it slowly to 86°F. (30°C.). Traditionally, stoneware pottery was used as the vessel, but enameled ware does well in its place. Add a scant demitasse spoonful of concentrated French rennet *(présure),* and keep the milk vessel in a cool spot at 55° to 65°F. (12.8° to 18.3°C.) to keep it from fermenting. The milk will curdle and curds and whey will separate. The curd must set till it is very firm, anywhere from twelve to twenty-four hours. Judging the right degree of firmness is a learned art. The curd must be firm enough to be placed in special draining cups that have holes on the sides and bottom. Old-fashioned, tinned draining cups were once the rule and are still preferred by many traditional French dairywomen, but the person who sent me my draining cups could only find the more modern plastic type. The curd must not slip through the holes, but it is important to start draining the curd as soon as possible. Evaluating the consistency of the curd and keeping it cool are delicate matters requiring repeated experience.

Fill the draining cups three-quarters full of firm curd and place the cups on a draining board. I use a stainless steel cake-cooling rack set over a basin. Allow the curd to drain for twelve hours. At that point, it should be firm enough to stand upright in your hand. Salt the curd by sprinkling all sides lightly with sea salt. This will produce a semi-hard-to-hard cheese. Then up-end the curd, reverse it, and place it back in the cup(s). Drain for twelve hours more. If you are interested in producing salt-free or very soft cheese, do not salt. However, I have found that unsalted curd tends to mold quickly inside the plastic cups here in the uncompromisingly damp Catskill climate. From this experience I would assume it is very hard to produce a soft cheese in a wet climate.

After twenty-four hours in the draining cups, remove the cheeses. The soft, farmer's-type cheese can be refrigerated or used immediately. In France, the ones that are to harden are laid upon a bed of straw, or suspended from hooks in a small rectangular cage of net or screening outdoors. I have found that bird's-eye diapers and the cake-cooling rack are adequate alternatives for airing. The cheeses should be outside at this stage and need protection from insects. *Demi-sec* will be cured and ready to eat in one to two weeks, depending upon the weather conditions. *Sec* takes three weeks or more.

If your curd has holes in it, too much curdling agent was used or the curd was left in the whey too long. It must go into the cups as soon as it is firm enough, but never earlier than eight hours from the start of the procedure. It will *look* firm long before it is actually ready to drain. Reserve a cup of the whey from the first batch to culture succeeding cheeses. The cupful is added in one or two tablespoon lots when the rennet is added, and all is gently stirred. I started my first batch by adding some shavings of purchased *chèvre* to the set curd.

Experience and adaptation to varying weather conditions are very important ingredients. In the opinions of Mmes. Roux and Litaudon, the kind of goat does not influence the cheese, but the goat's diet does, and so does thunder-showery summer weather. In March and April, Mme. Litaudon brings her mixture into the kitchen so that it will be warm enough. When the weather gets hot, humid, and heavy, the milk vessel is moved outside to some shaded, cool spot.

These cheeses are known in our household as the "holey-cup cheeses" and are well worth the little extra watchfulness they require. The French holey cups can be reproduced by a good stoneware potter.They come in various sizes, are smooth and straightsided, and have holes pierced through the bottom and sides. It is interesting that these country cheesemakers have also evolved nonpressing methods for handling goat milk curd.

Main Dish Recipes

Now that the goat gourmet has a storeroom full of diverse cheese treasures, we can return to main-dish use of goat produce. Over the summer season, one of the most savory uses of the goat and the garden is to prepare the vegetarian answer to the quiche Lorraine:

Vegetarian Quiche Lorraine

2 onions, sliced
4 large Swiss chard leaves, chopped
½ cup boiling water
½ cup vegetable oil (don't use olive oil)
2 cups whole wheat pastry flour
½ teaspoon salt
¼ cup toasted sesame seeds
1 cup scalded goat milk, cooled to lukewarm
3 eggs, beaten
1 cup hard, grated goat cheese (you may use soft cheese)
dash of nutmeg
½ teaspoon tarragon

Sauté onions and chard until tender. Pour into a colander to cool and drain. In a bowl, pour the boiling water over the vegetable oil and blend well. Cut in the whole wheat pastry flour (you may need slightly more than 2 cups) and salt. Mix thoroughly and add sesame seeds. Roll the dough into a ball and chill thoroughly for 30 minutes or more.

Blend the milk, eggs, and hard cheese (if you're using soft cheese, layer it with the vegetables when making up the pie) and the spices.

Roll out the dough and place it in a 10-inch pie pan. Place the drained and sautéed vegetables in the crust. If you are using soft cheese, sprinkle it over the vegetables. Pour the milk-egg mixture over all. Bake at 350°F. (176.7°C.) for 30 minutes.

Many variations on this basic theme are possible. Add alone or in combination:

—cut and drained fresh red ripe tomatoes
—small cubes of tofu
—thin layer of puréed and cooked beans
—½ cup of cooked and drained chickpeas
—1 cup cooked, drained, cubed eggplant or
  summer squash
—1 cup steamed broccoli
—1 cup sautéed and drained sweet peppers

This dish is easily adapted and expanded to fit increases in amount of garden produce or quantity need. It is very flexible since it is tasty served hot or cold. When chilling, sogginess can be avoided by taking extra care to drain all vegetable additions well.

Another excellent way to use home-produced milk, eggs and goat cheeses is to prepare soufflés. The dairy goat soufflé is the envy of all those who delight in French cookery because it achieves unusual lightness and taste.

Dairy Goat Soufflé

*3 tablespoons butter or
  vegetable oil
3 tablespoons flour
1 pint goat milk
¼ teaspoon sea salt
dash of nutmeg
4 to 6 eggs, separated*

*1 cup hard, grated goat
  cheese
1 cup assorted chopped,
  sautéed, drained
  vegetables: broccoli,
  green peppers, onions,
  tomatoes, squashes*

Make a cream sauce by melting the butter in a saucepan over a medium flame and then adding the flour, blending with a whip. When well blended, slowly add the goat milk, stirring to prevent lumps from forming. Add sea salt and nutmeg. Add beaten yolks to the cream sauce, stirring well.

Add cheese and vegetables, remove from heat. Beat the egg whites until they form stiff peaks and slowly fold them into the cream sauce. Pour the entire combination into a buttered soufflé baker of 2-quart capacity. Bake at 350°F. (176.7°C.) for 50 to 60 minutes until the soufflé has risen and the top is puffy and golden. Rush to the table with the soufflé and eat it quickly, for it collapses when it hits the cooler air outside the oven.

Omelettes made with the brined cheese and spinach or Swiss chard are piquant and similar to Greek spinach pies. Brined cheeses substitute ideally in all recipes calling for *feta* cheese. They also work well in dishes calling for mozzarella and in Mexican-style burritos or enchiladas that call for melting cheese.

Since little or no meat is consumed in our household there is a definite need for a source of vitamin B_{12} in our diet. We get an adequate supply by incorporating dairy products in our diets. Cholesterol is not a primary worry because we're all very active and because we are adequately supplied with unsaturated fatty acids from vegetable oil sources. After our first winter in the rough mountain climate hereabouts we had to work in more fatty foods than we were used to eating in the city. The richness of goat's milk products has supplied that need.

Desserts are always eagerly anticipated when they are old-fashioned custards or a puff of a goat soufflé:

Honey-Goat Custard

4 eggs, beaten
2 cups goat milk, scalded
   and cooled to lukewarm
¼ teaspoon sea salt
1 tablespoon whole wheat
   pastry flour

2 ounces honey
½ teaspoon vanilla
   extract (or scald milk
   with a vanilla bean in
   it)

Blend all ingredients and then pour into lightly greased baking dish. (Individual cups may be used but baking time will be shorter.) Dust top with ground nutmeg. Set dish in a shallow pan of water and bake in a 350°F. (176.7°C.) oven for 50 to 60 minutes. Test for firmness by inserting a knife. Custard is set when the blade comes out clear and clean. Add ½ cup desiccated, unsweetened coconut for a fine variation. Reduce the honey to 1 ounce because the coconut is naturally quite sweet. Adding 1 cup of cooked and well-drained winter squash or pumpkin makes an unusual honey-custard.

Rice Pudding Custard

2 cups cooked or leftover brown rice
½ cup raisins, currants, or dried apricots and ½ cup chopped almonds
3 beaten eggs
2 cups scalded goat milk, cooled to lukewarm

2 tablespoons honey
¼ teaspoon sea salt
¼ teaspoon cinnamon
¼ teaspoon ground cloves
¼ teaspoon nutmeg
1 tablespoon whole wheat flour

Place cooked rice in lightly greased baking dish. Add dried fruit and nuts and mix well. Blend remaining ingredients and pour over rice mixture. Set dish in shallow pan of water and bake at 350°F. (176.7°C.) for 30 to 40 minutes till custard has set. A layer of sliced apples lining the bottom of the dish works well too.

The French dessert cake known as *quatre-quarts* can be turned into a natural food treat:

Natural Quatre-Quarts

4 eggs, separated
4 ounces honey
4 ounces goat butter

½ to 1 cup whole wheat flour
jam

Combine egg yolks with honey, then add goat butter and mix well. Mix in flour. Beat the egg whites until they form stiff peaks and fold slowly into the first mixture. Pour the batter slowly into a shallow, buttered ring-type mold. I usu-

ally do this in two stages, placing several teaspoons of homemade jam atop the first pouring of the batter, then adding the rest of the batter to seal in the jam.

Bake at 375°F. (190.6°C.) for 50 to 60 minutes. After removing from the oven, allow to cool for 30 minutes before releasing from pan. Half a cup of raisins and 2 tablespoons of lemon juice produce a good-tasting variation.

Cheesemaking seems to lead naturally into custards and soufflés but there are many dishes that utilize milk as the major ingredient and depend little upon cheese. Soups fall into this category, and seem to be increasingly popular. Creamed soups make good use of seasonal fresh or frozen home-grown produce and are highly nutritious and digestible made with goat's milk. Broccoli, lettuce and cauliflower are spring or cool weather vegetables that work well in soups. Tomatoes, zucchini, and celery are the major summer soup vegetables. Potatoes, pumpkin, cabbage, and root vegetables are the mainstays of winter soups.

Creamed Broccoli Soup

2 cups fresh, chopped
   broccoli
1 chopped onion, including
   green top
3 tablespoons oil
3 tablespoons whole wheat
   flour

1 quart fresh goat milk
2 tablespoons butter,
   yogurt, or cream
1 teaspoon curry powder

Sauté the broccoli and onion in 1 tablespoon of the oil. Add a few tablespoons water and cover tightly. While the vegetables steam, heat remaining oil in a saucepan over moderate heat. Stir whole wheat flour into the oil, then slowly stir in 1 pint milk. When sauce is thickened, add it to the steaming vegetables, and allow mixture to simmer until thick. Put 1 cup of goat milk in blender, add half the vegetable mixture and purée. Repeat with remaining milk and vegetables. Return all the soup to a saucepan and slowly reheat. Add 2 tablespoons of butter, or yogurt, or cream, and 1 teaspoon curry powder. This soup serves 4 heartily, and can be eaten hot or cold.

Cauliflower Cream

1½ pints water
1 head cauliflower
1 tablespoon butter or oil
1 tablespoon whole wheat
   flour

1 pint goat milk
2 tablespoons butter,
   yogurt, or cream
salt, nutmeg to taste

Boil water and then cut and add entire head of cauliflower. Cover tightly and cook until the cauliflower is soft. Prepare a light cream sauce as described above with the butter, flour, and milk. Purée cauliflower with part of its steaming water and add to cream sauce mixture. Put entire soup through a blender, if smooth texture is desired, adding more cooking water if mixture is too thick. Return soup to heat, add butter, yogurt, or cream, salt and a dash of nutmeg. Depending upon the size of the cauliflower, this soup serves 5 to 7 persons.

Cream of Green Soup

2 large heads romaine
   lettuce
3 onions, including green
   tops
3 tablespoons oil
1 tablespoon whole wheat
   flour

1 teaspoon curry
1 teaspoon tarragon
½ cup water
2 cups goat milk

Wash, drain, and cut lettuce finely. Chop onion finely, sauté until transparent, and add lettuce. Sauté briefly until lettuce wilts. Add whole wheat flour, curry, tarragon, and ½ cup of water. Simmer 3 minutes. Purée ingredients in an electric blender. Return to flame and slowly stir in goat milk. Heat until well blended. Serves 4 to 6.

This soup can be made with wild greens substituted for half the lettuce. Sorrel alone makes an excellent variation. Serve the sorrel soup cold with a goat yogurt garnish. One cup of fresh green peas are a delicious addition; add to the soup when it is returned to the flame for reheating.

153

Cream of Tomato Soup

1 small onion, sliced
3 tablespoons oil
1 quart goat milk

4 tablespoons whole wheat
  flour
4 cups cooked tomatoes
1 teaspoon honey

Sauté onion in oil until transparent. Add milk and simmer. Mix flour with a few tablespoons of water and then add to the milk and onion. Cook over low heat for 15 minutes until thickened. Purée cooked tomatoes in blender with honey and a dash of salt. Sieve to remove seeds and then add tomatoes to the thickened milk. Stir well and reheat. Pour soup into serving bowls containing a teaspoon of butter or grated goat cheese. Add chopped fresh basil and chives to bowls. Serves 6 to 8.

Cream of Zucchini Soup

2 pounds diced zucchini
2 onions, minced
2 cloves garlic, minced
3 tablespoons oil

3 tablespoons whole wheat
  flour
1 quart goat milk
¼ teaspoon oregano
¼ teaspoon thyme

Steam vegetables till soft in ½ cup of water. Prepare cream sauce as for Creamed Broccoli Soup. Add cream sauce to steamed vegetables, stirring thoroughly. Add herbs and gently reheat together. Serves 4.

Cream of Celery Soup

5 cups diced fresh celery
2 onions, minced
2 potatoes, diced
1 clove garlic, minced

3 tablespoons oil
3 tablespoons whole wheat
  flour
1½ quarts goat milk

Sauté all vegetables in oil in a large kettle. When they are soft and transparent, stir in flour until well coated. Slowly add 1 pint of milk, constantly stirring. Remove from heat source. Blend mixture in blender, adding more milk as necessary. Return puréed mixture to soup kettle. Reheat to just below scalding temperature. Top with chopped scallions or parsley. Serves 6 to 8.

Pumpkin Soup

1 onion, diced and steamed
  with'
1 small pumpkin or winter
  squash
2½ cups water (use water
  from steaming)

2 cups goat milk
¼ teaspoon powdered
  cloves
¼ teaspoon nutmeg
¼ teaspoon ginger
cooked brown rice

Blend vegetables, water, and milk in electric blender. Return to heat and add spices. Serve over 2 to 3 tablespoons cooked brown rice in individual bowls. Sometimes I reheat this soup in a large, hollowed-out pumpkin in a slow oven, and serve it from the shell at the table.

Cream of Roots Soup

1 large turnip, thinly sliced
2 carrots, diced
1 parsnip, thinly sliced
4 diced potatoes

1 diced onion
1½ cups water
2 cups goat milk

Bring vegetables to boil in the water in a tightly lidded soup kettle, reduce the flame and allow to simmer until the vegetables are soft. Mash the vegetables, then remove half and purée them in a blender. Return to the kettle and slowly add goat milk. Cook over low heat until thick. Serves 4 to 6.

Country Velvet Potato Soup

3 onions, thinly sliced
3 tablespoons oil
3 pounds potatoes, thinly
  sliced

2 cups water
1 quart goat milk

Sauté onion in oil, till very soft. Boil potatoes in water till very soft. Blend 2 vegetables in blender or mash through food mill, using the cooking water and some milk as needed. Season to taste with salt and return to stove for slow reheating. Add remaining milk and cook on low heat for 10 more minutes. Add chopped chives or scallions just before serving. Serves 8.

Cream soups can form the core of a meal when accompanied with a whole grain bread and a salad of raw vegetables. Root vegetables that are steamed to doneness, then mixed and baked briefly in the oven topped with four to eight ounces of goat milk and a cheese topping also provide a protein-rich core casserole. In the summer, tomatoes, eggplants, and squashes can be turned into fine casseroles by simply slicing the available vegetables, layering them with slices of fresh goat cheese and green onion tops, and baking in a hot oven for ten to fifteen minutes. When the surplus gets out of hand, I dice up the tomatoes, eggplant, squash, and onions, sauté them until tender, and can the mixture as *ratatouille* by adding four tablespoons vinegar, one-half teaspoon salt, and garlic, basil, and oregano to taste per quart of mixture. *Ratatouille* forms a fine winter meal either as a casserole base or as a filling for whole wheat crêpes layered with melted goat cheese.

Beverages

My final use for goat milk is as a beverage. One of the most refreshing and varied, and easiest things to do with raw goat milk is to infuse herbs into it. The mint family tastes particularly good in milk. Elder flowers, lemonbalm, and bergamot are also good infused in raw milk. Fennel and aniseed in milk are old European favorites. The peppermint milk is a pick-up; the seed milks and bergamot are soothing and relaxing aides to digestion. Elder flower milk is refreshing and has the reputation of helping to purify the blood. Heat milk to *below* scalding, pour it over the herbs, and allow it to stand ten minutes with a cover or flat dish over the cup. The infusion may be sweetened with a little honey for children's palates. Starting winter mornings with a warm honey-nutmeg-goat milk combination cheers the grayest days.

Goat Meat

Goat meat is known in English as "chevon." In my youth, local Italian butchers used this term to their American customers to refer only to meat from animals older than kid stage. Italian and Greek sources I've queried distinguish only between young, entirely milkfed kid meat and goat meat from animals older than two months. Some writers refer to several stages: the milkfed kid of four to six weeks of age, male kids slaughtered soon after birth (seven to twelve pounds), young wethers of six to eight months age, and old animals being culled from the herd.[12,13]

It is not economically advantageous to raise wethers, the castrated bucks, for meat purposes today unless you are raising your own hay and grains. Even then, dairy-type animals do not develop the optimal chunky, blocky meat-type body. When the keeper is compelled to buy feed and hay out, the buck will require $80 to $100 in support outlays and does not return a lot of dressed-out meat. On a straight dollar basis, it is probably more viable to put those dollars and surplus milk into rearing piglets to meat, for the feed-meat conversion ratio is more favorable in pork raising. For the meat-eating household, I would recommend raising surplus buck kids only to the three-to-six-week age-range and feeding them exclusively on goat's milk. This will yield a superior, tender, white-fleshed meat, and requires no additional or extended cash support. Kid meat is relatively fat free as compared to lamb. It is delicious, and without the stringy, sometimes tough character of chevon. This approach also puts the least pressure upon the keeper when considerations of stall space are a major concern.

Direct, guided experience is vital before undertaking any type of homestead butchering. I suggest you make contact with experienced local farmers, butchers, and hunters, and that if interested in butchering, you be present at and participate in several slaughtering and dressing sessions before going ahead singlehandedly. The two methods I have seen my local free-lance butcher use for kid slaughter are employing a hammer and awl-like instrument to deliver a blow to the center front of the head, and using a small hand gun, both of which resulted in a clean kill and instantaneous death. Both approaches depend on practice and an unflinching attitude. If you are not prepared to take butchering on as a routine chore, it is worth the few dollars to have young buck kids slaughtered, bled, cooled, and properly dressed out by an experienced farmhand, hunter, or game butcher. Fall is the busy season in my region for such people. In the spring, when kidding season is at its height, it is usually possible to find someone available for these jobs at a minimal cost. The simplest way for the keeper who does it at home, and the cheapest way when engaging a butcher, is to dress the kid out in quarters. Each quarter forms the core of a meal, packs easily into the freezer, and requires little extra labor or fancy carving skill.

When undertaking a slaughter at home, do not feed the kid(s) for a day, and then totally remove them from the eyesight and earshot of the dam and herd. After killing the animal with either a gunshot to the head or a blow directed as described above, cut the jugular vein cleanly and hang the animal head down to bleed out. If you improvise and hang the animal from a structural member of a building or from a tree branch, make sure to select a high point, so that dogs, roaming raccoons, or other predators have no access. Skinning is done by slitting the carcass pelt deep, from the hind legs up to the throat, and making ancillary cuts along the inner sides of the legs. Cut and loosen the pelt from the anal area. Some people tie off the colon at this point to prevent any spillage of the intestinal material. Once you have pulled the pelt away from the anus, work to separate the thin, furry skin from the legs. The pelt can usually be pulled away up to the ears once the leg and anal parts of the pelt are freed. The pelt is released by working the fingers up between it and the actual flesh. After the kid is skinned, cut open a belly slit to eviscerate the animal. Watch others do this a few times on deer or lambs before setting out to remove the intestines and inside organs. Getting the inner organs out intact and clean takes initial guidance followed by practice. Hang the carcass up in a cool place or refrigerator locker overnight, then quarter and freeze it or use it.

A traditional kid dish is prepared as follows:

Capretti lattanti alla Nonna

Rub down a half or a quarter kid well with olive oil. Superficially slash meat in 5 or 6 places and insert fresh garlic slivers (2 to 3 cloves is a reasonable total to use). Sprinkle lightly with rosemary or oregano, and place on rack in roasting pan. Preheat oven to 400°F. (204.4°C.) and place kid in oven with a light covering of aluminum foil over it. Roast at high heat for 10 to 15 minutes, then reduce temperature to 350°F. (176.7°C.) and roast uncovered till done, basting occasionally with a mixture of 3 parts white wine to 1 part olive oil. Allow 20 minutes roasting time per pound of meat. In Mediterranean countries this is the traditional roast served for the spring holidays.

If you decide to cull an older animal, call upon a farming neighbor to give you a hand, or someone with substantial deer hunting and dressing experience. An older goat's meat is more tender and tasty after a period of marination. A marinade I've mixed is made up of:

Tamari Marinade

½ cup tamari (soy sauce)
1 cup white wine
½ cup olive oil

chopped green onion, garlic, and rosemary to taste

Chunk up the chevon to be served and soak it in this mixture overnight or for a minimum of 4 hours.

Tender curried chevon can be made as follows:

Curried Chevon

1 cup goat yogurt
2 cups water
1 cup marinade from marinating meat
2 tablespoons fine ground whole wheat flour
3 tablespoons olive oil
½ teaspoon mustard seed
½ teaspoon cumin seed
1 teaspoon fenugreek seed
2 whole cloves

½ teaspoon sea salt
1 teaspoon tumeric
1 teaspoon pulverized coriander seed
¼ teaspoon powdered ginger or 1 teaspoon fresh grated ginger
2 dried cayenne peppers (depends upon individual preferences)

Blend yogurt, water, marinade, and flour, and let stand. Heat oil in heavy cast-iron skillet and then add first 3 seeds. When seeds start to pop about, add liquid mixture to skillet. Reduce heat to lowest possible point and then add salt and other spices. Simmer for 10 minutes, then add chunks of chevon and simmer for 15 minutes more. Serve with rice and an accompanying salad of cucumbers dressed with lemon and yogurt.

CHAPTER 12

THE GOAT IN THE GARDEN

Long before the dawn of my goatkeeping adventures, I had picked out, turned and planted a 35-foot-by-30-foot garden. The site was predetermined — it was the only place that was open to the sun, while most of the grounds were planted to diverse, and well-grown trees and ornamental shrubs such as lilac, forsythia, and honeysuckle. There were also areas overrun by dense, tangled growth of sumac, and there were patchy outbreaks of wild blackberries, ragweed, and Chinese elm. I did not want to disrupt earlier plantings and could hardly cut my way into the scrub jungles. So the garden was planted on the sunny spot. The following spring revealed that the garden had been located within the foundation and rubble of a burned-out house. The garden began to yield a peculiar crop of bed springs, lengths of wire, galvanized pipe, and other assorted old implements of metal, stone, and ceramic. Large blocks of the original foundation's stones were heaved up over the winter and surfaced in the garden in the spring. The soil turned out to be a few inches of fill. That first garden yielded, but it was a happy combination of beginner's luck and a thin layer of soil that had lain fallow for two years except for growing crops of weeds that, in dying, added some organic matter to the plot. It became top priority to build up the soil over the bedsprings and pipe fittings. Our first carrots had come out of the soil looking like ginger roots. We figured that we needed to build eight inches of soil to grow any root crops at all. The first year we hauled sheep, cow, and horse manure from the nearest farms (which were not near at all) and from defunct chicken and dairy barns. Rotted hay was added to the manures, and all our homemade compost from kitchen garbage, brush cuttings, and weed clippings went on the garden. We succeeded in adding almost two inches of composted material over that initial year, but it took great expenditures of human and fuel energy to locate and haul in such quantities of matter. I was already looking into goats for other reasons and was being rapidly won over to them in a personal way. Searching for a reasonable source of homegrown manure had us turning over the idea of keeping goats and chickens. The immediate imperative of garden soil improvement reinforced the process.

Goat Manure

When the first goats came to my place, I had decided to make the garden top priority and to let methane fantasies simmer in the background. I went into a period of diligent research on the subject of goat manure — its nitrogen, phosphorous, and potash breakdown and its fertilizing yield. The upshot of the research on the subject was a final round zero. Though the dairy goat was the earliest productive animal domesticated for human use, probably five or six thousand years ago, the popular homesteading literature has no statistical analyses or even rule-of-thumb information on how much manure one could expect a dairy goat to produce in a year. Horses are supposed to produce nine tons a year, and a cow about eleven tons. Formulae based upon how much an animal is fed multiplied by a preordained factor (2.1 for a horse, 1.8 for a sheep, 3.8 for a cow) plus the weight of bedding utilized, will not work for goats on the browse.[14] The impossibility of collecting and computing the manures of browsing animals seems to be the stumbling block. There are some analyses of goat fecal matter, but the diet fed modern milk goats is very different from the diet fed milk goats in the past. Organic gardeners and homestead authors are either silent, vague, or contradictory on the subject of goat manure as garden fertilizer. I ended my bookish researches and went farm to farm in my area asking goatkeepers what they were doing with the manures produced in the course of their everyday operations.

It surprised me to find that more than half the goatkeepers I spoke with did not utilize the manure produced by their animals. Two keepers stockpiled the manure outside their barns and used it to dress shrubs and other ornamentals. Six keepers used it as a vegetable garden fertilizer; two used nothing else. One forty-eight-year veteran composted his chicken barn litter and the goat bedding manure together and used no other materials on his gardens. Many keepers simply kept goats to show or as a hobby, but did not consider themselves gardeners or food producers. I also heard several stories of inexperienced gardeners putting goat-barn bedding directly on their gardens because they had read that goat manure was by nature "cold" and "precomposted," as opposed to "hot" manures such as horse manure. The results included destruction of the plants and root burning.

All fresh, urine-soaked goat bedding is "hot" because the urine contains from one-half to two-thirds of the total nitrogen in the manure mix. The dung pellets, euphemistically known as "goat berries," contain almost all the phosphate component, and perhaps can be classed as the "cold" manure. However, when most keepers muck out a stall, they clean out a hot mix of urine-soaked dung pellets and partially broken down bedding material. Although placing this mass directly on the garden and working it into the soil insures that most of the nitrogen and potash will go into the soil, it definitely will do in tender plants.

Manuring the Garden

My first method of putting the goats in the garden had nothing at all to do with manurial analyses. One bright November morning, I was struck with the innovative idea of turning the goats into the fall garden remnants. They could graze amid the rubble of broccoli stems and corn stalks, clean up the stubby, resistant leavings on the site, deposit their pellets — and we would all mutually benefit. The goats were delighted with the new twist in their regimen. Pasture had been drying up all around them and they had begun a seasonal turn to evergreen needles and bark. Once in the garden, they chewed and chomped away at the succulent garden crop residues with relish and delight. They did an excellent job of clearing. Little did I know that I had created an indelible imprint in their minds.

All the next spring I had to struggle to lead the goats past the garden gate. Mackenzie would stop, plant all four feet firmly and not yield an inch once she was near the gate. She would stand there craning her neck and gazing into the wire mesh squares, perhaps seeing in her mind's eye the rows full once more of cabbage stumps and cauliflower leaves. At night I would dream of the goats escaping the pasture enclosure and flying over the garden fence to gobble up the tender shoots of the lettuces, turnips, and peas that were, in reality, coming along fast in bursts of fine spring energy. As my morning conflicts with Mackenzie lengthened, aggravated by the necessity of leading the goats past the garden, dreams continued over the passing nights. I vowed never again to graze the goats among the stumpy leavings of the fall garden.

Though I was resolved, the goats were still entertaining other ideas. One summer afternoon, I returned from some errands and saw high-flying Eleanor clear the garden fence and in lightning gestures gobble up the heads of the small, green sunflowers. She was heading back down the row for the leafage when I caught up with her and led her out of the garden to the ignominy and constraint of a tether. The final resolution of the situation came only with the construction of a fenced corridor allowing direct access from the barn to the pasture so that the goats never again had to saunter past those delectable garden grazing grounds.

Please learn from our experiences and put your goats in the garden via the manures and composts they yield. We now have over eight inches of true garden-quality, life-supportive, organically enriched soil in most of the garden. We have developed several basic practices and have some additional suggestions for people just starting out. Average barn manure is a mix of excrement and urine-soaked bedding, sometimes given a theoretical value of 0.5-0.25-0.5 (referring respectively to nitrogen, phosphoric acid and potash).

The USDA Handbook of 1957, *Soil,* states "the expense of applying 100 pounds of 10-5-10 fertilizer is much less than that of applying a ton of manure. On soils in good tilth, the returns from an equivalent amount of fertilizer usually will be greater. . . . It can be concluded that at least in northern latitudes manure does not provide all of the plant nutrients needed and fails to provide any plant nutrients or growth supplements that cannot be applied by artificial fertilizers." The statement was made twenty years ago when the 100 pounds of 10-5-10 was $2.50 the hundredweight. We now have to face the net energy costs of producing and applying that chemical fertilizer and many farmers and gardeners are rapidly finding it more cost effective and clearly environmentally safer to return to utilization of manures. There is also an underlying assumption in the quoted statement that soils will simply go on being in good tilth. Long experience of chemical fertilizers used exclusively demonstrates the exact opposite — soils that are compacted, exhausted, and mined. Repeated dependence on artificial fertilizers alone reduces tilth. Even proponents of chemical fertilization have had to grudgingly recommend the incorporation of organic matter into the soils because of the humus supplied. Good water retention, proper aeration, and favorable microorganism activity are the benefits achieved by employing manures.

Goat barn manure is used here in several forms. Stalls are cleaned on an average of every three to five months. This produces a deep, partially composted litter. The stalls of kids and doelings are cleaned at four- or five-month intervals, the milkers' stalls at twelve-week intervals. The bedding material is composed of wasted feed hay which the goats obligingly pull out of the feeders and scatter over the stall floors as they pick and choose choice tidbits for dinner. Probably a pound or two gets scattered on the floors daily when the goats are confined to the barn in deep winter. Litter and excrement do not accumulate as rapidly when the goats are on pasture as when they are always stalled or closely yarded.

At the end of the gardening season, just prior to the onset of truly cold weather in November, all the stalls are mucked out and the wetter, urine-soaked manure is hauled by wheelbarrow or pickup truck to the garden and spread on the plot to be turned under, time and weather permitting. This is fresh, hot manure. There is some nitrogen loss because the material is exposed to the air and the ammonia formed in the early stages of fermentation and decomposition is volatile. However, the loss is much less than would be incurred if rain leached the nutrients out of a storage pile. Part of this load also goes out onto the compost pile nearest the garden and is laid down in alternating layers with chicken barn litter, household garbage, garden crop residues, and wood ash. Less-soaked litter goes directly on the *Rosa rugosa* bushes as a fertilizing winter mulch.

In February, the stalls of the milkers are mucked out and the material is put into a compost pile layered with some chicken manure. When the milkers' stalls are cleaned during cold weather, a shallow layer of manure is left in the stall to provide the new bedding with a working population of microorganisms. If the weather is very cold, the stall is left with a third of its litter, so that the animal still has the benefit of the warmth released from the composting layer of litter beneath her. Initially, we hauled every bit of barn manure to the garden so the soil could increase quantitatively. Now, the concern is to improve the soil qualitatively. Manure is introduced primarily as finished compost except for the fall layer used to tuck in the land for hard weather. Two or three piles are working; one of them is targeted for spring or early summer application, usually the pile from the prior

November. With material in abundance, we have had no need for specialized shredding equipment to speed up the process of composting. As stalls are cleaned in the spring and summer seasons, the litter goes into compost piles that get worked over thoroughly by a small flock of free-ranging chickens. The birds scratch and shred the material into very finely broken-down matter. We have resolved the old dilemma of whether to use fresh or rotted manures by using some of each. In this way, the nitrogen loss incurred by allowing manure to rot and age is offset by fall application of fresh manure. In addition, I top-dress all young, leafing vegetables with a good home brew of goat berry tea.

Goat berry tea is made by digging deep into a milker's stall and half-filling a burlap feed sack with goat berries and bedding. Suspend the bag in a fifty-five-gallon drum and fill with water. Cover the brewing vessel tightly with a lid to prevent insect penetration. If you do not have a lid, float a thin slick of recycled motor oil on the top of the water to seal it off from flies and mosquitos. (Be sure not to mix the oil with the water you pour on plants.) Allow to steep for three to four weeks. The resultant "brew" is a manure tea that provides an instant nitrogen boost to plants. It is used after plants have six to eight real leaves. When feeding it to transplants, wait until the young ones have been in the ground two weeks or more and have conquered transplant trauma. The tea is an excellent and inexpensive boost for corn in its early stages and encourages fast, deep green growth. I do alternate feedings of the fledgling vegetables, watering with goat berry tea one week, and then feeding or foliar-spraying with a dilute seaweed emulsion two or three weeks later. I have found this regimen a safe, economical, and effective alternative to using nitrogen boosters such as cottonseed meal or bloodmeal. These are both costly and objectionable, because cotton is apt to be heavily sprayed and some of the practices in the slaughtering industries involve dangerous chemicals. People come to our place with sacks to scoop up batches of goat berries to brew up as dressing teas for their roses and prize ornamentals.

Since manure is often described as being deficient due to leaching or atmospheric loss and requires supplementation to release the full value of its components, I have also taken to feeding my garden soil strong solutions of comfrey tea, described by Lawrence Hills as "comfrey liquid manure," as well as the diluted seaweed emulsions. Analysis with my home testing kit shows no major deficiencies. The need for liming has been almost nil, with soil pH remaining quite steady year to year on fresh and composted goat manure plus the supplementation practices outlined above.

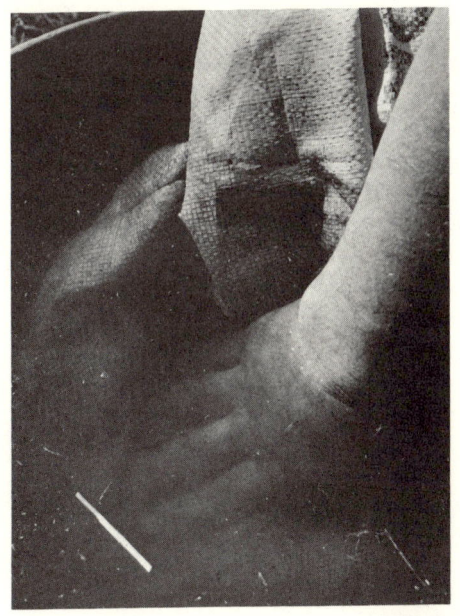

Manure tea is easily made by scooping some manure, fresh or composted, into an old feed sack, then immersing it for at least several hours in a bucket of water. The resulting brew is a great fertilizer.

Our Garden, Based on Goats

The goat in the garden contributes a vital constituent to the life cycle. Usually there is complete concentration on milk productivity when the dairy goat is described in books and periodicals. It is in the context of the small, household-oriented herd that a truly accurate accounting of productivity can be made. Ninety percent of our household's garden fertilization needs are met by goat manure, manure teas and goat-manure-based composts. In a good year, our fickle weather permitting, one-third of our food supply is grown in the now expanded 35-foot-by-70-foot garden and ancillary 15-foot-by-35-foot corn patch. In 1973, the garden yielded $490 worth of vegetables. The expanded garden produces over a thousand dollars' worth of vegetables a year, and most of the expanded plot has not yet achieved a soil depth of eight inches. This invisible income is nontaxable and consumes little or no fossil fuel resources in production, preparation, transportation, processing, or marketing. It actually accrues value, for this activity enriches the land and makes available to the household food of a nutritionally higher standard than the food currently available in commercial markets.

The value of the goat's manure cannot be quantified, but it is an ideal fertilizer for the small-scale, organically oriented household gardener. The materials yielded are adequate fertilizers produced in reasonable tonnage. Little or no cash expenditure is required for supplements. Until we bought our pickup truck, we carted the manure out of the barn to the garden by wheelbarrow — between eight and fifteen cartloads per stall. No more costly or demanding piece of equipment than a rented tiller has been employed to process all this material. It is either composted or tilled under directly. An assortment of inexpensive handtools — spading fork, pitchforks, shovels, and hoes — are the only other tools needed to produce fertilizer for our yearly vegetables. This aspect of the dairy goat's productivity should not be glossed over, for it is of definite ecological and financial importance. Alas, we have no Advanced Registry system for recording and acknowledging manurial production.

Large reserves of compost and an unending supply of goat manure have enabled us to deal with the shallow soil problem presented by our garden site. The reserves have made possible a raised-bed type of garden. We have developed a variant of the French intensive raised-bed method that utilizes raised rows. Our technique owes its derivation to an experimental cross between French and traditional Chinese methodology. The rows are deeply dug and filled with compost, with the soil layer replaced atop the compost, raising the original row, with its top flattened, to a height of eight or ten inches above ground level. The rows are easier to weed or water than a twenty-four-inch bed. We garnered stubby, stunted root vegetables until we tried this approach. In the first year we used raised rows, we produced beets that reached a size of two pounds each and bushels of turnips, in addition to our first recognizable carrots and parsnips. For persons who want to make the transition from a summer seasonal garden to a year-round, household sustenance garden, increased root crop production is essential. Root vegetables are the mainstay storage vegetables, and it is critical to household economy to produce them so that home-produced food is available throughout the winter. However, raised rows would also be of great use to persons with limited garden space, for the enriched rows permit intensive succession cropping, and consequent lengthening of the growing season if planted to light-feeding, frost-hardy crops of greens, such as Swiss chard, mustards, and lettuces.

The success of the raised rows led us to employ a layer of fresh goat manure as a bottom layer in a cold frame. It was topped with finished compost mixed with garden soil. The same layers were used in starter hotbeds for raising plants. The cold frames and hotbeds now extend the season by producing greens all but two months of the year. We also start hundreds of seedlings in the beds. If we had had to haul all manures from other farms, we might never have gotten to these projects. An accretion-type, solarized lean-to greenhouse will make possible the production of fresh greens throughout January and February. We will build the glazed walls and structure, but the enriched growing medium will be provided by the goats.

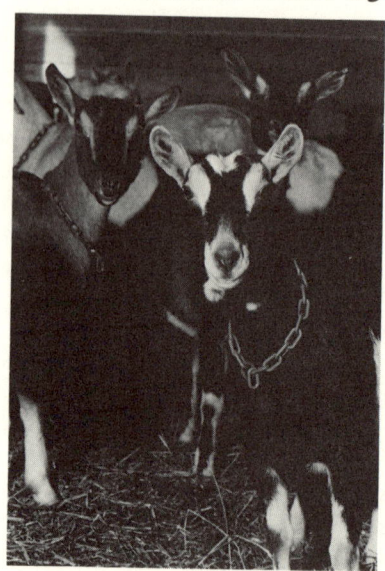

CHAPTER 13

GOAT ECONOMICS

People always ask, "How much does it cost to keep a goat?" There is no set answer to such a question. It is possible to compute how much it costs to keep a particular goat for one year in a specific locality in relation to a chosen system of management, but it simply is not possible to say, off the top of one's head, what it costs to keep goats in general. There's just too much variation from barn to barn, geographic region to region, from animal to animal, to generalize in any meaningful way. If the questioner wants to know how much cash outlay was required for a particular herd over a specific period, the question is easily answered. Recording purchase prices of hay, grain, and other supplies is easy. If feed is produced on the farm, purchase prices prevailing in the area can be used and noted against the costs of labor, fuel, maintenance and depreciation of equipment, land taxes, and so forth, which must be figured to calculate the costs of on-farm support of a herd. This chapter will include an approximate analysis of a household goat economy. It will also give the reader a glimpse at current modes of operation on commercial-scale goat dairies and specialized breeding farms. In my eyes, goat farming can be broadly divided into three often-overlapping categories: the household herd, the commercial dairy, and the breeding enterprise.

The Family Herd

Small family-oriented dairies have received much attention in the homestead literature of the last five years. The economics of household dairying are very positive as long as the herd is kept to a truly manageable size — three or four does in milk — and there is pasturage available. I have even found that labor costs are covered for an operation this size by homestead-generated, goat-based income.

Before surveying the situation of the more commercially oriented herds, it is interesting to take a more detailed look at the contemporary backyard-type goat enterprise. One of my interests from the outset was to chart the labor input required by keeping a few goats, and to check at the end of a year to see whether the tending labor was at all compensated by the year's goat-related income. The steps involved in doing this are:

— Recording all hay, grain and miscellaneous expenses.
— Recording milk yield, based upon weigh-in each morning and night.
— Recording any income derived from keeping goats: stock sales, milk or cheese sales, etc.

Records turn out to be fun and record keeping an easily entrenched habit once you really get going with it. Matching up the yield and income sheets against the expenditures list gives the keeper a quick and vivid picture. The labor involved in tending a small number of goats— to me a small herd is four animals maximum, and no bucks — averages out to between one and two hours daily. (The margin allows for efficient or inefficient barn arrangements and differences in numbers.) The figure includes the daily milkings; the grooming and hoof-maintenance sessions; barn cleaning and maintenance; fence installation, mending and checking; hauling of hay, grain, and other supplies; kid-rearing; and sanitation and general processing and handling of milk equipment. Tallied generously the labor required runs between 700 and 1,000 hours yearly.

Articles that give estimates of the upkeep costs of a milk goat often omit any labor cost and many small keepers, when queried about this omission in their own computations, have no idea of paying themselves for their labors. The argument made by the small keeper who omits to calculate labor runs something to the effect that the "goats are kind of a hobby, and what other folks would pay to go out and be entertained goes to support the home entertainment provided by the goats." Other statements I have encountered included, "It's something I like doing; getting paid for it would make it like a job," and "I'm satisfied just being able to cover my cash outlays," or "My table dairy products come free, so that covers labor." Most small-scale keepers are quite content to cover all the maintenance expenses and not worry about labor compensation because they either keep their goats to improve the quality of dairy products in their diets or to improve the quality of milk goat stock, as an avocation. Many small owners do not produce or market enough goat products or stock to pay themselves, and have made specific decisions against getting into bigger or more commercial ventures. So there is little information available on the total cost of small-scale goatkeeping. Most of what follows is based on my own records, goat journals of the last five years, comparative interviewing, and reading of materials from different regions in the United States.

Over the last six years, I have arrived at some general figures for cash outlay and labor. It must be noted that my figures are based upon a management design that includes six to eight months of daily outside pasturage intake, and no special labor such as that for showing grooming, schooling, or traveling. In 1972, every hundred pounds of 14 percent commercial dairy ration cost an average of $4.15 in our locale. Hay of medium feed quality cost approximately eighty cents per bale. At this writing, 1977, retail prices have just about doubled for both hay and grain. At the current average of eight cents a pound for feed grain and $1.50 per bale for average quality, mixed legume-grass or hay, my milkers have been consuming ninety-six dollars' worth of purchased feed materials per year, plus all they can forage in browsing. (The average milker here utilizes about 750 pounds of grain a year and up to 1,000 pounds of hay.) The hay consumption varies enormously with the quality of the available hay and the severity or mildness of the winter months. The grain figure allows about 2 pounds per milker, maximum, and also varies considerably in relation to the doe's production curve and outside pasturage conditions. Other costs to be considered include allowances for miscellaneous purchases such as salt, minerals, wheat-germ oil, possible veterinary expenses, utilization of electric power for fencing and lighting, water supply, and advertising costs. In all, a mature milking doe is supported by a cash outlay of $125 in a region with high overhead and by a labor outlay of approximately two hours daily.

What can the new keeper expect for this input? The productive modern milk goat covers all her costs. A good milker averages between 1,500 and 1,800 pounds of milk over ten months, roughly 750 to 900 quarts per year. In my area, the retail value of goat milk averages $3 and up per gallon. Computing at the retail value — what a family would have to pay at the supermarket — the dairy goat hypothetically returns over $600 yearly in milk alone. When you are buying all her feed and hay, supporting her pastureland via taxes and paying for her other miscellaneous needs, her cash support requires a maximum of fifty cents daily, and her yield can return over $2 daily to the keepers. If one could tend four goats with a yearly input of 1,000 labor hours, and each returned more than $600 in income, the keeper could earn a projected $2 an hour after cash expenditures. A cow dairy farmer of my acquaintance told me that in a good year, he nets less than $1 an hour for his work over 365 days.

It would seem that goat farming should pay, for even the smallest of keepers, but here we come to the heart of the matter. Most keepers do not begin to cover their labor, primarily because they cannot use or market all the milk they produce, or cannot get their herd to produce well enough to bring in a return. It is only in rare or long-established situations that goatkeepers market all their milk at a favorable price. The realities of milk marketing are usually fraught with legal restrictions, seasonal ups and downs and prevailing low prices. Often the small, rural keeper responds by using surplus goat milk to rear other more easily sold foodstuffs such as veal calves. Though the paper economics of small goatkeeping are overwhelmingly favorable, the actualities are that the keeper can cover labor and cash expenses only after several years, given a well-researched market and foundation stock that produces over 1,500 pounds yearly and a healthy kid or two every twelve months. My records and conversations with owners of large-scale goat dairies have also made it clear that goat dairying on a larger scale can definitely provide a living, as long as initial capitalization and running overhead costs are not too high.

The Commercial Dairy

The countryside rolls out flat north of Madison, Wisconsin. Dairy barns and storage silos spring up from the cleared expanses, often the only landmarks in the landscape for many miles. The huge twin stacks of Columbia County's giant coal-fired electric generating plant dwarf all the barns and other buildings on the approach to Portage. The land maintains this planed-down character until one comes upon Caledonia, nestled in a bend of the Wisconsin River. Here the land is hilly, forested, old and rich. On 121 acres of prime Caledonia slope, Harvey Considine raises dairy goats, his own grain and hay to supply them, the family's beef, and its fruit and vegetable supply. Considine, who describes himself as an organic farmer, is interested in maintaining a high standard of soundness, "with a view toward long-run land enrichment and livestock improvement." He says, "I'm not a nut in any one direction," just an organic farmer returning what he can to the land, interested in balance, and farming with a long view, with an eye to dairy goat improvement.

In the heartland of Wisconsin's cow dairy country there are several full-scale commercial goat dairies producing Grade A milk primarily for urban markets. Considine has been marketing goat milk, raw goat milk cheeses, and purebred stock of all the major breeds, producing them on organically managed land, since 1946. Considine started out as a hobbyist for about thirteen years, then managed a large goat dairy, and finally purchased the operation, and, most importantly, its well-established market, urban Chicago. The purchase of the rights to

Harvey Considine feeding newborn kids using the milk from the mother goats.

Daniel Considine of Sunshine Farms milks his goats by hand during the winter. It keeps his goats used to handling.

Jane Considine feeding young goats in the kid house.

Goats being housed in the main barn at Sunshine Farms.

the Chicago goat milk market was fundamental to making Diamond Farm a viable enterprise. Supporting that is Harvey Considine's life-long experience in farming and dairying, more than thirty-two years of experience with dairy goats. At one time he was milking 450 head and farming 600 acres. Now fifty-one, he describes himself as "semi-retired." His present herd numbers approximately 120 animals. He recently sold his extensive raw goat milk cheesemaking business and part of his herd to a neighbor and close friend, Rev. Paul F. Ashbrook, who Considine knows will maintain the quality built by Diamond Farm.

As one of the few people in America to have made a living in commercial goat dairying, Harvey Considine has depended on income from three sources: sales of purebred stock, income earned at shows, and wholesale milk and cheese sales. He feels that the dairy that tries to make a go of it on milk sales alone will have some tough times. Building a good purebred herd of diverse breeds will increase productivity for the dairy and will also generate substantial income through the sale of kids. Purebreds, he believes, will always realize a solid income for the dairy farmer; grades will not show a similar return and have little or no show income potential.

In his early years of supplying the Chicago goat milk market, Considine found that demand did not keep up with his herd's seasonal production peaks, so he began to make cheese with surplus milk. He started very small, but eventually employed a master cheesemaker, purchased a walk-in cooler, and developed the market entirely from scratch to the point where he carried a $30,000 cheese inventory to go through a winter. The cheese enterprise grew to such proportions that he recently sold it so that he could devote more time to dairy goat classification, gardening, land and herd management, and writing.

Cheese production and marketing was a long, slow learning process. Decisions had to made regarding what types of cheese to produce, what distribution channels to develop, what packaging and labeling to adopt, and how to price the products. Over the years Diamond Farm has produced Colby, Cheddar, Swiss, and Muenster cheeses, and sold salted and unsalted cheeses. Eventually operations were simplified, prepacked, prelabeled, standardized units that elimi-

Stanchion setup at Sunshine Farms. Goats enter in the rear and do not have to jump up or down from the stanchions. This helps to prevent leg injuries, and it saves the backs of the milkers who don't have to bend as much and don't have to lift heavy milk containers from the floor.

nated selling large blocks of cheese were instituted, and certain types of cheeses that kept poorly were discontinued. Raw milk cheese requires informed and sensitive handling from producer to distributor to consumer with constant attention to proper refrigeration and stock rotation. At the time Considine started, he was able to begin small and move slowly, expanding gradually; now he estimates a small cheese business would require $6,000 to $10,000 initial capital. Purchase of a milk truck and walk-in storage cooler would probably run in the $25,000 range. All such figures are rough estimates and would vary with location and other such factors.

Harvey Considine's son David farms more than 200 acres over the mountain from his father. David's Breezy Hill Farm is a diversified family farm that includes a commercial orchard, approximately seventy acres a year in hay and a goat dairy herd of around 150 animals. He has not inherited land, stock, or capital, but was given a part of his father's interest in the Chicago milk market. Readers should understand that wholesale prices for goat milk average half of the retail price per gallon. If this is related roughly to the cow dairyman's nine

The kid house at Sunshine Farms, complete with playpen. The goats are outside for at least one hour daily, even on the coldest days, and they seem to love it.

Kids are fed milk using this multi-nippled feeder.

One of two large pens in which goats are housed at Breezy Hill Farm.

David Considine feeding his young goats.

dollars per hundredweight of milk sold wholesale, the goat dairy farmer might command a price of between twelve dollars and twenty-one dollars per hundredweight. It is clear that the commercial dairy farmer needs the income derived from the sale of stock. David Considine currently runs a large debt, as do most dairy farmers, but he feels he will make it in goat dairying in the long run. If he had not grown up with good dairy goats and come into part of his father's Chicago trade, he is not sure whether he could make it pay. He does not keep an exclusively purebred herd at present because the cost of purebreds was too expensive when he started out. Although he feels that there is a place for good grades in commercial goat dairying, he is moving toward an all-purebred herd because income of stock sales is so much higher for purebreds.

Management practices are very similar at Diamond Farm and Breezy Hill Farm. The goats are kept in a common stall arrangement

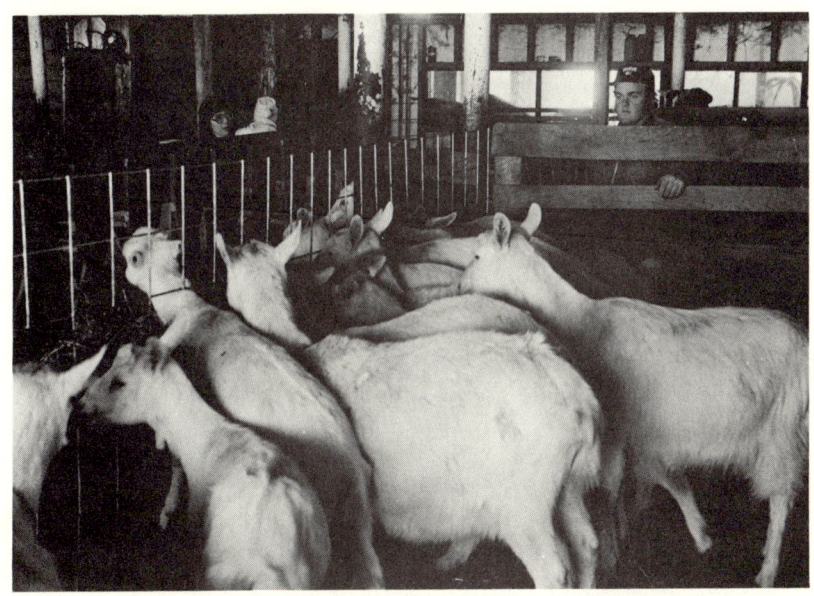

The Saanen is David Considine's favorite breed.

The Rev. Paul Ashbrook of Idelmar Dairy Goat Farm with one of his French Alpine goats.

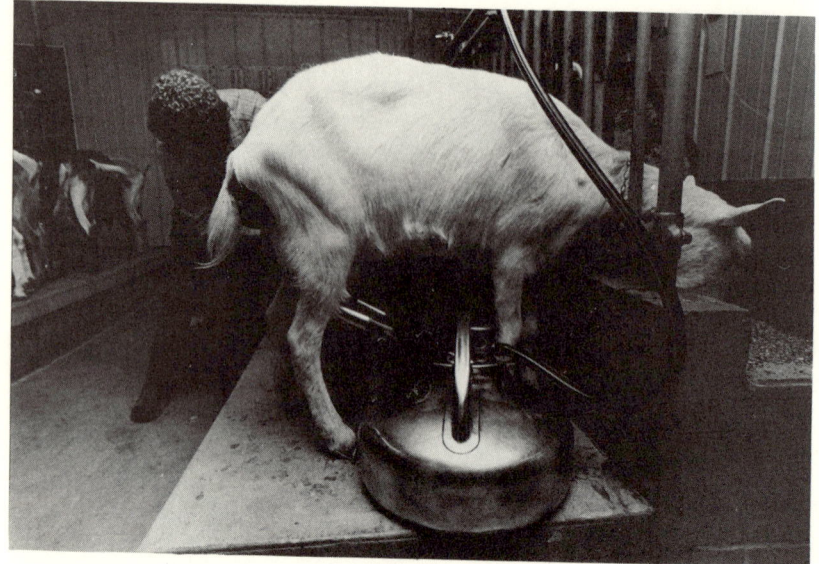

Milking machine in use at Idelmar. One hundred fifteen goats can be milked in about three hours. Two machines are used, and twenty goats are stanchioned at a time.

with access to an exercise yard. The goats are not pastured since both farmers feel it is hard to maintain an even milk quality when goats eat just whatever is available, but the goats are not kept in heated barns or coddled in any way. David feeds whatever chopped fresh greens are seasonally available and raises choice alfalfa hay on his own acreage for cold weather feeding. All grains fed are raised on the farm and both farmers feed a simple corn-oat ration with molasses added. Minerals and salt are available on a free-choice basis. Grain feeding averages between one and two pounds per milker daily, with seasonal production curve adjustments. Harvey Considine gets additional protein and greenery into his does by pelletizing his early alfalfa with some corn, linseed meal and minerals, in addition to feeding high quality legume hay free-choice. Harvey Considine finds that the pellets make for less wastage of valuable legume hay. David has also utilized pellets but now has the hay resources to feed his alfalfa straight, without the added expense of bringing his crops to the mill for pelletizing.

Both farmers describe themselves as organic in farming orientation and general land management practice. Harvey Considine composts the goat manure right in the barn with the addition of Radiant-blend microbial activator, powdered rock phosphate, and lime as needed. Broken-down litter is cleaned out only a few times yearly, and put back directly on the land. He estimates that his plowing, cultivating, and fertilizing costs come in well below those of the average chemically oriented dairy farmer. He works his land on a five-year rotation pattern.

Harvey and David's one exception to organic farming practice is resorted to when weather prevents timely access and working of the soil for the corn crop. Their usual method of corn raising is a program of seeding, blind cultivation, cross-dragging, and a good early cultivation after the corn is up. However, many times in a wet spring, Con-

When the machine is full, it is emptied through a strainer into a bulk milk tank.

sidine can't work his soil on critical days. In these circumstances, he has utilized one-half to one pound of Atrazine per acre banded lightly along the corn rows. David Considine's land shares much of the same soil character as his father's but was not as carefully managed in the past, and he has to work harder at general soil improvement.

A third Considine goat dairy is located east of Portage, and is run by Daniel and Stephen Considine and Daniel's wife, Jane. The management practices at Sunshine Farm are somewhat different from those at Diamond Farm and Breezy Hill. The goats are pastured for part of the year. Sunshine Farm has no La Manchas in the herd. And the owners describe themselves as absolutely organic. Income derives from stock sales, cheese sales, show-ring income, and the sale of bulk milk into the Madison and Milwaukee markets, which they have developed themselves. They also sell large numbers of buck kids to the meat market.

The partners are working over 400 acres — "more than we really need," they say — and putting in long hours and extra capital on a soil improvement program. The farm started out on rented land and the partners carry a large debt and are working hard at paying off

At Idelmar, hay mangers are located on both sides of a long walkway, which allows them to be refilled easily.

their own land, which at the same time needs much input to produce a healthy and organically enriched soil. Daniel feels it is possible to run a commercial goat dairy and make it on milk sales alone and is encountering more people who are doing it. Absolutely essential, he feels, to such a venture is good market research and extra care put into promotion and market development. Sunshine's approach is to have several sources of income and to stress their quality which is supported by the organic approach.

All three farms have milking parlors and equipment that meets local standards for production of Grade A milk. The goats are milked at two runs of metal stanchions that each accommodate ten or twelve head; groups of twenty or twenty-four does are milked at once. The parlor has cement floors and milkstands that can be easily swabbed down. Harvey Considine utilizes milking machines but hand finishes each milking doe year-round. Daniel and Stephen hand milk in the winter months when production is light. They have found that during these months hand milking is as time and energy efficient as machine milking when one has to clean underutilized milking machines. David tries to do as much hand milking as is productively feasible, feeling that machines have contributed enormously to the mastitis problems so common today in cow dairying.

All concur that there is a future in commercial goat dairying, even though their approaches differ. A well-developed market and long hard work, through the intial years of large-scale debt, are basic, as is an extensive background in goat farming and breeding.

The Breeding Enterprise

The breeding-focused goat farm has always been a more common type of enterprise than the large-scale dairy farm. Frank Holder of Pleasant Valley, New York, has been keeping purebred Nubians for more than twelve years. The focus of Poverty Hill Farm's program has been to produce the highest yielding herd of purebred Nubians in the east. Holder's long-term goals have been to breed selectively to increase the quantitative output of the purebred Nubian, and to extend and maintain a high-yielding lactation, tackling one of the basic areas of Nubian stock improvement. He does not show his animals for a variety of reasons, the primary of which is that he feels that productivity drops overall when does are on the road. He refuses to sacrifice his long-range goals for show income. Holder purchases all hay and grain, and pastures his goats when weather permits. Housing is of the common stall type. Most of his labor time is spent in hand-rearing stock, in basic chores, and in promoting and marketing his stock.

Human energy is concentrated on the breeding and raising of stock rather than on the production of support crops. Income is primarily based on stock sales with the sale of milk-fed livestock such as veal or pigs as back-up income. Sales are based upon continuous DHIR testing, consistent and well-designed advertising and the slow building of a herd reputation as the highest-producing purebred Nubian herd in the region. Frank Holder usually has enough from stock sales to live on.

Another example of a farm concentrating on stock improvement is that of Gerald Lehmann of Pine Plains, New York, who has been involved in some form of dairying most of his life but who now raises purebred Toggenburgs as a passionate avocation. He is interested in a long-lived, efficient and productive dairy goat, and sees the Toggenburg breed as the one breed that has progressed closest to his ideal. This herd is supported by the income derived from full-time, off-the-farm employment, and all tending and chores are done in Lehmann's spare time. The herd is kept necessarily small, usually numbering no more than a dozen animals, breed bucks included. This is still considerably larger than a household-oriented herd, and enables the owner to see improvements and changes over several generations and among differing lines. Kids are raised on their dams till six weeks of age, and then sold, usually to local 4-H'ers. Hay and feed are purchased. Income from stock sales is not large but usually covers cash expenditures. The owner's goal is stock improvement and not labor compensation. However, he has evolved a system that is efficient and requires less input than is spent on the other farm described above, having eliminated large numbers, crop raising, hand rearing of stock, and consistent sales promotion.

Keys to Success

By considering these differing goat farming enterprises, the reader can gain some ideas on the alternate approaches and implications of breeding or commercial dairying commitments. My household came up with still another, hybrid, solution. No one here was particularly interested in showing goats, and no one had the background, resources or interest to run a full-time, 150-head goat dairy. We arrived at a common decision to produce milk primarily for household purposes, to sell our surpluses in the form of cheese, and to keep our operations small and in the black, just as if we were running a miniaturized goat dairy farm. Additional goals are upgrading our stock and closing the circle of homestead productivity by providing fertilization for our sustenance gardening.

From what I have seen locally, goat ventures do not seem to work out positively for many people. Goals are not realized, and frustration appears over and over again, often for the same reasons. The first pattern one sees in failing goat ventures is that of overextension. Too many animals are accumulated in the early years. Reluctance to part with spring kids builds the size of herds fast. All the pedigree assurances in the world will not compensate for poor management. And poor management seems to follow directly upon overwhelming numbers. A swamped goatkeeper will inevitably wind up with a mediocre or poor herd, get discouraged, and then get out of goatkeeping in short order.

The second common pattern one sees is that of unrealistic market projections. Lack of research into the legal and demographic limits of a particular area can lead to large and early losses. Many people keeping goats on a backyard or homestead basis find they have a steady and reaonable demand for their stock or milk products. From that set of personal experiences, they jump to the conclusion that their area needs or could support a large, commercial-scale dairy. Careful, extended groundwork and analysis of local conditions must be undertaken before making such leaps.

A third frequent pattern is that of entering dairying or stock breeding with a sentimental outlook as opposed to a pragmatic one when it comes to cultivating livestock productivity. People with this tendency regard milk goats as adorable pets rather than producers. The second and third pattern are often encountered together and make a tightly reinforcing net.

A fourth reason for the brief careers of many goatkeepers, small and large, is out and out undercapitalization. Many persons fall in love with goats and desire to make goatkeeping a vocation rather than a hobby. They elect to set themselves up as dairy operators or professional breeders. Often, however, they lack the land, capital, credit, or extensive prior experience requisite to make a go of it. Either of these undertakings requires accumulated resources and a clear financial idea of the long-term nature of the venture.

Finally, errors in stock selection seem to be very common. Many people simply do not do enough comparative looking and analysis before purchasing. Although there are extensive materials about selection criteria available for study, not many folks use them unless they have had substantial prior livestock experience. If you are starting out for the first time, doing your goat homework (as described in chapter two) is necessary.

Setting limits and goals, financial and other, is very helpful for making good selections and so is the ability to be flexible in approaching stock acquisition. If you have your heart set on Nubians to supply milk for your table, but the nearest purebred Nubian buck available for service is 250 miles away, it may be more rewarding to acquire the nearest well-bred stock, even if you end up with two Toggenburgs. If your primary goal is to develop stock of a particular breed rather than to supply your household with dairy products, then you will want immediately to acquire the best purebred stock of that specific breed. Clarifying one's real intent is the key to gratifying goatkeeping.

Once we had decided upon our intentions and goals, we chose a breed that suited these needs, and set the following priorities for the venture: (a) supplying the home table, (b) fertilizing the garden, and (c) generating income via sale of stock and surplus in the form of cheese. We then culled or acquired as the situation demanded. By putting care, knowledge, and affection into goat management, the keeper comes to perceive and enter into the oldest yet most vital experience of humanity, the nurturing cycle, and to reap its benefits and to contribute to its process. In actuality, the goat economy can be microcosmic; in implicaton, it is universal.

ARTIFICIAL INSEMINATION

The fundamental argument for employing artificial insemination in dairy goat breeding is to improve genetic capacity for production. In dairy cattle, milk productivity has nearly doubled in the years between 1950 and 1974, largely due to genetic inheritance improvement, itself based on artificial insemination (A.I.). More than 50 percent of America's cows are enrolled in an A.I. program and almost every herd utilizes it to some extent.

A.I. is based on using semen collected only from superior sires capable of transmitting the characteristics of high dairy productivity. In this way a high frequency of occurrence of desired genes nudges out the genes for low productivity and poor conformation. This is true grading-up. The semen collected from chosen sires is frozen and can be stored and shipped over long distances in liquid nitrogen, making it available to small and large herds, far and near. Most small herds now are usually forced by circumstances into using the nearest bucks, whatever their abilities to improve productivity over the long run. Dr. H. A. Herman argues strongly for the increased use of A.I. among all types of herds, suggesting that owners learn to be their own technicians, or employ local cattle inseminators or veterinarians, and that owners of small herds pool their resources to buy or rent a liquid nitrogen storage tank (about $250) and in these ways bring the genetic capabilities of the best sires in the country to herds everywhere.[31]

Active in this campaign to promote increased use of A.I. is Andrew Purcella of Coquille, Oregon, who has written, taught, and extensively demonstrated methodology for bringing A.I. technique to herd owners all across the United States.

Basically, A.I. is achieved by introducing semen from a storage ampule via a long plastic straw called a pipette, as deeply interuterine or as deeply cervical as possible. Usually more than one pipette is needed. Technique is vitally important and is explained in a step-by-step reprint of a Purcella article that appeared in the August 1974 *Dairy Goat Journal,* a must for anyone wishing to pursue selective breeding of dairy goats utilizing A.I. Purcella's best results have been from inseminating does toward the end of the standing heat period (when she would stand for a buck) through the next twelve hours. This contradicts Israeli and French research on the subject, which recommends insemination at the earliest possible time in the heat for highest conception rates. Attending an A.I. demonstration clinic is the next step I would recommend. Goatkeepers will often find that the local cattle inseminators are too busy, too inexperienced with goats, or unwilling to inseminate milk goats, and that the task will fall to the herd owner.

Though the productivity-increase evidence for employment of A.I. is indisputable, it is not within easy reach of the homestead herd owner for reasons of geographical isolation, expense, or technical training. Getting owners to pool money for the storage tank is very feasible, but storage sets up many timing losses due to back-and-forth travel to pick up ampules. It is also difficult to gather a group of people with the financial and time resources and also the passionate commitment to genetic improvement needed to make a group A.I. undertaking worthwhile and functional. For large and small enterprises to truly have access to A.I. and the resultant productivity increases, small herd owners may have to pressure their land-grant colleges and local clubs and national registries to research and subsidize A.I. for dairy goats. It is a job beyond the scale and resources of many small owners. Yet it is the small herd owner who stands to benefit greatly in the long run, for A.I. could bring the lines of the best animals into the most modest herds.

NOTES

1. Harold B. Gotaas, *Composting* (Switzerland: World Health Organization, 1956). Excellent technical monograph on methane production.

2. "Rules for Registration and Recordation," American Dairy Goat Association. Available from ADGA, P.O. Box 186, Spindale, NC 28160. Cf. Countess de Saint Seine, "The Real French Alpines," *Dairy Goat Journal* 52 (November 1974) 18-19. "Here no registration is allowed for animals who have not produced at least 1,500 pounds in an adult lactation. And practically no male is sold for breeding whose dam has not given more than 2,000 pounds."

3. USDA, "Improvement of Milk Goats," *Yearbook of Agriculture, 1937: Better Plants and Animals,* pp. 1294-1313; photograph, p. 1298.

4. W. F. Brannon, "The Dairy Goat," Cornell Extension Bulletin no: 1160 (1967), p. 18.

5. Review of the ADGA *Handbooks* for 1974-77.

6. David Mackenzie, *Goat Husbandry,* rev. ed. (London: Faber and Faber, 1970), p. 207.

7. *Handbook* (ADGA, 1973), pp. 74-75. The all-breed milk record holder is Puritan Jon's Jennifer, II (1960), 5,750 pounds in 305 days; Puritan Jon's Janista holds the Toggenburg butterfat record with 202.5 pounds in 305 days.

8. Swiss Alpines or *Oberhasli* are currently experiencing a revival in the U.S.A. The identity of the breed, known for its excellent milk quality and good temperament, was almost lost through interbreeding with other alpines, but numbers are now increasing. For information, contact Oberhasli Breeders of America, Rt. 2 Box 59B-1, Cumberland, VA 23040.

9. F. B. Morrison, *Feeds and Feeding,* 20th ed. (Claremont, Ontario, Canada: Morrison Publishing Co., 1961), pp. 803-4, "In general the same feeds and the same care and management that are successful with dairy cows and sheep are suitable for milk goats."

10. USDA, "Principal Poisonous Plants," *Yearbook of Agriculture, 1942,* pp. 356-57. Lois Hetherington, *All About Goats* (Great Britain: Farming Press Ltd., 1977), pp. 96 and 146.

11. The best material to fill this gap is to be found in *Hoard's Dairyman* articles on forage crops and in the series, now ongoing, on grasses and legume plants in *United Caprine News.* See the bibliography for the addresses of these publications.

12. C. E. Spaulding, *A Veterinary Guide for Animal Owners* (Emmaus, Pa.: Rodale Press, 1976), p. 78, and J. Belanger, *Raising Milk Goats the Modern Way* (Charlotte, Vt.: Garden Way Publishing, 1975), p. 62.

13. Jill Salmon, *The Goatkeeper's Guide* (N. Pomfret, Vt.: David & Charles, 1976), pp. 50-52.

14. Gene Logsdon, *The Gardener's Guide to Better Soil* (Emmaus, Pa.: Rodale Press, 1975); USDA *Yearbook of Agriculture, 1957: Soil.* Cf. Sir Albert Howard, *An Agricultural Testament* (Emmaus, Pa.: Rodale Press, 1976); F. H. King, *Farmers of Forty Centuries* (Emmaus, Pa.: Rodale Press, 1973); and Gotaas.

15. Most goat books organize the "disease and accident" chapter on

the alphabetical pattern for reference convenience. Authors following this pattern start with Pegler at the end of the nineteenth century and continue right up to the present including Belanger, Mackenzie, Spaulding, de Baïracli-Levy, Downing, Shields, and Hetherington. My decision to depart from this body of authority is based upon student feedback and the experiences of many people recently starting out who have most often wanted to know what commonly to expect.

16. Hetherington, p. 116; Mackenzie, p. 267, differs and gives a wide range of 70 to 95 for the pulse and 20 to 24 for respirations.

17. Jeffrey Williams, "Internal Parasitism in Goats," *Dairy Goat Journal* 55 (May 1977), pp. 22–30, and Dale R. Nelson, "Health Problems in Dairy Goats," *Dairy Goat Journal* 55 (February 1977), pp. 3–11.

18. Williams, p. 24.

19. Ibid., p. 29.

20. Ibid., p. 30.

21. Judy Sinclair, "Coccidiosis in Goats," *Country Women,* issue 23, pp. 35–37.

22. Ibid., p. 37.

23. Spaulding, pp. 97–100.

24. Mackenzie, p. 229, implicates parasites as causes of abortion in early pregnancy. Belanger, p. 82, picks this up. Hetherington, p. 120, cites the microbial agents *Brucella melitensis* and *Spirochoet* but emphasizes the rarity of occurrence.

25. See note 10 above.

26. Sherry Thomas and Jeanne Tetrault eds., *Country Women: A Handbook for the New Farmer* (Garden City, N.Y.: Anchor Books, 1976), pp. 355–58.

27. F. A. Wright and E. T. Oleskie, "Here's an Update on Ketosis, " *Hoard's Dairyman* 123 (January 10, 1978), pp. 14 and 60. Spaulding, pp. 119–20. Tetrault and Thomas, pp. 363–64. Wright and Oleskie do not agree with the Tetrault-Thomas molasses recommendation.

28. Spaulding, pp. 82–88; Thomas and Tetrault, pp. 261–66; Mackenzie, pp. 232–38.

29. Goat pox is a minor disease characterized by small, crusty, blister- and pimple-type skin eruptions. It is contagious and relatively harmless, usually immunizing the animals the first time around. Scabs can be a nuisance but the entire course of the outbreak is usually brief. Don't pick the scabs off because the runny matter from the blister seems to spread the condition. I have seen it localized primarily around the udder. Usage of goldenseal infusion for bathing and comfrey ointment can be resorted to in more severe cases. Milk affected goats last and wash your own hands thoroughly.

30. M. J. Ojo, "Caprine Pneumonia," *Dairy Goat Journal,* January 1978, pp. 12 ff. reviews the confusing scientific literature on contagious caprine pneumonia and states that control is difficult because "adequate research has not been carried out," p. 27. Pneumonia resulting from incompetent drenching and from parasite inroads seems to be occurring only rarely now. Today, stress, lowered resistance, and sudden changes seem to be frequently implicated in pneumonia discussions.

31. Harry A. Herman, "Use A.I. to Boost Herd Income," *Dairy Goat Journal,* July 1974. Reprint available from *DGJ.*

BIBLIOGRAPHY

Books

de Baïracli-Levy, Juliette. *Herbal Handbook for Farm and Stable.* Emmaus, Pa.: Rodale Press, 1975.
> An alternate healing manual for livestock and other domestic animals that is of invaluable aid to those seeking a nondisruptive approach to handling problems. It is based upon deeply experienced, long-term empiricism, and utilizes materials natural to the European and Mediterranean environment, most of which are found in the Americas too.

Belanger, Jerry. *Raising Milk Goats the Modern Way.* Charlotte, Vt.: Garden Way Pub., 1975.

Colby, Byrin E. *et al. Dairy Goats, Breeding, Feeding and Management.* Originally published by University of Massachusetts with the Massachusetts Co-op Extension, 1966; now available from the ADGA.

The best, least costly starter's booklet that adequately covers all the basics. Recommended for the person starting to toy with the idea of goatkeeping.

Downing, Elisabeth. *Keeping Goats*. Great Britain: Pelham Books, 1976.

Gotaas, Harold B. *Composting: Sanitary Disposal and Reclamation of Organic Wastes,* No. 31. Switzerland: Monograph Series: World Health Organization, 1956.
Technical but very comprehensive treatment of composting, rural waste utilization, and appropriate-scale technologies to meet these needs. It has unusually factual and well documented and illustrated material on methane generation.

Heidrich, H. J., and Renk, W. *Diseases of the Mammary Glands of Domestic Animals*. Translated by L. W. Van den Heaver. Philadelphia: W. B. Saunders, 1967.
A specialized veterinary text that has a pathology focus, but it treats the milk goat as a serious dairy animal. Good, sometimes unsettling visual material on udder diseases, including goat udder photos.

Hetherington, Lois, and the T.V. Vet. *All About Goats*. Great Britain: Farming Press Ltd., 1977.

Jeffrey, H. E. *Goats*. Great Britain: Cassell, 1970.

Mackenzie, David. *Goat Husbandry*. Rev. ed. London: Faber and Faber, 1970.
Still the best and most comprehensive goat book, worldwide in perspective but decidedly British in focus, orientation, and style. Some of the material is dated, but most remains appropriate to the small-scale keeper. Mackenzie's approach and opinions still raise the fires of controversy among professionals and amateurs alike.

Owen, Nancy Lee. *The Illustrated Standard of the Dairy Goat*. Rev. ed. Scottsdale, Ariz.: Dairy Goat Journal Publications, 1977.

Pegler, H. S. Holmes. *The Book of the Goat*. 9th ed. London: Thorsons Pub. Ltd. Reprinted by American Supply House, 1965.
Still quite interesting as the progenitor of multitudes of modern goat books and for its accounts of the formation of the Anglo-Nubian breed.

Salmon, Jill. *The Goatkeeper's Guide.* N. Pomfret, Vt.: David & Charles, 1976.

Shields, Joan and Harry. *The Modern Dairy Goat.* C. Arthur Pearson, Ltd., 1949. Reprinted by Tiger Press.

Spaulding, C. E. *A Veterinary Guide for Animal Owners.* Emmaus, Pa.: Rodale Press, 1976.
> Excellent general reference book for the homestead shelf. Spaulding relies primarily upon western veterinary therapies and drugs but is conscientious and conservative in his goatkeeping recommendations.

Thomas, S. and Tetrault, J. *Country Women: A Handbook for the New Farmer.* New York: Doubleday, 1976.

Periodicals

Acres, U.S.A., 10227 East Sixty-first St., Raytown, MO 64133.
> Not specifically focused on dairying, *Acres* runs occasional features on ecologically oriented dairy farmers. Styling itself a "voice for eco-agriculture," it gives you an idea of what the alternate minority of large-scale farmers are experimenting with these days in large-scale composting, seed treatment, and foliar spraying that avoids toxic chemicals.

Country Women, Box 51, Albion, CA 95410.
> Feminist country magazine that carries a wide selection of practical articles. There was a very heavy emphasis on goatkeeping in the first few years. Back issues are worth seeking out. Some of this early material has appeared in expanded form in the anthology edited by Tetrault and others (see above).

Countryside and Small Stock Journal, Waterloo, WI 53594.
> Worthwhile and consistently informative general periodical with good material on dairy goats.

Dairy Goat Journal, Box 1908, Scottsdale, AZ 85252.
> Devoted exclusively to dairy goats, the *Journal* is the best ongoing source for general goat-dairying information, and is of particular interest to the breeder, commercial dairy person, and other professionals involved with showing goats.

Hoard's Dairyman, Fort Atkinson, WI 53538.
> This is the dairy industry magazine, and it is usually packed

with high-quality and up-to-date information, 99.9 percent of it about cow-dairying. Its orientation is that of contemporary American agribusiness, and the inexpensive subscription rate is made possible by the extensive equipment and chemical industry advertising. When read carefully, with an eye to its politics and support money, it can be a useful tool. Cheap and biweekly.

United Caprine News, Box 1042, Abilene, TX 79604.
A newcomer to the periodical literature, *UCN* promises to evolve into a good generalist publication covering widely differing aspects of goatkeeping.

RESOURCES

Registry Organizations

American Dairy Goat Society, Box 186, Spindale, NC 28160.

American Goat Society, 1606 Colorado St., Manhattan, KS 66502.

Canadian Goat Society, Canadian National Live Stock Records, Holly Lane, Ottawa, ON K1V 7P2.

Suppliers

American Supply House, Box 114, Columbia, MO 65201.
General livestock equipment.

Dairy Association Co., Lyndonville, VT 05851.
For "bag balm."

Green Mountain Herb Co., Box 2369, Boulder, CO 80302.
For echinacea, goldenseal root.

Charles Hansen Laboratory, 9015 W. Maple St., Milwaukee, WI
53214.
　　Animal and "vegetable rennet."

Henry Doubleday Research Association, Bocking, Braintree Essex,
England.
　　Comfrey ointment.

Hoegger Supply Co., Box 490099, College Park, GA 30349.
　　General livestock supplies.

Horan-Lally Co., Ltd., Rexdale, Ontario.
　　Canadian source for rennet.

Nasco, 901 Janesville Ave., Fort Atkinson, WI 53538.
Nasco West, 1524 Princeton Ave., Modesto, CA 95352.
　　Nationwide supplier of livestock equipment.

H. W. Naylor Co., Morris, NY 13808, Dept. H-11A.
　　"Blukote."

Omaha Vaccine, 2900 O St., Omaha, NB 68107.
　　Veterinary and pharmaceutical supplies.

Red Mountain, Box 758, Glenwood Springs, CO 81601.
　　"Incredible cheese curdler" vegetable rennet.

San Francisco Herb Co., Box 40604, San Francisco, CA 94140.
　　Bulk and specialty herbs.

INDEX

A Abomasum, 27
Abortion, 116
Abscess, 113-15
    antibiotic for, 114
    inoculation against, 113
    lancing of, 114
Accidents, 116-17
Advanced registry doe, definition
    of, 5
Afterbirth. *See* Placenta
Age, for first breeding, 45
Alfalfa hay, 77
American Dairy Goat Association,
    4, 19
American Goat Society, 4, 19
Americans, of breed, definition of,
    4
Amprobium, for coccidiosis, 111
Anemia, worms and, 110
Aniseed, added to milk, 156
Anthelmintic, definition of, 109
Antibiotics, use of
    for abscess, 114
    during drying off period, 53
    for mastitis, 123
Appetite
    loss of
        coccidiosis and, 111
        enterotoxemia and, 118
        ketosis and, 119
        milk fever and, 120
        worms and, 67, 81, 109
    of young kids, 62-63
Artichokes, Jerusalem, as fodder,
    75
Artificial insemination, 42, 189-90
Azalea, toxicity of, 75, 117
B Bag. *See* Udder
Balloons, to practice milking, 33

Barbed wire, for fencing, 99, 116
Bedding, during winter, 86-87
Beginner goatkeeper, advice for,
    13-14
Bergamot, added to·milk, 156
Berries. *See* Manure
Birthing. *See* Kidding
Bloat, from overeating, 117-18
Blood
    in milk, 31, 123
    in stool, 112
Blood poisoning, after abortion,
    116
Blood sugar deficiency disease, in
    pregnant does, 119
Blue cheese, 146
Body temperature, 84, 108
    rumen and, 27
Bolus, to administer worming
    compound, 109
Bones, pelvic, appearance of,
    before kidding, 60
Books, value of, 13
Bottles, baby, to nurse kid(s), 58
Bran, for pregnant does, 51, 80,
    120
Brassicas, as fodder, 75
Bread, recipe for, 134
Breathing
    patterns before kidding, 60-61
    rate of, 108
Breech presentations, 120-22
Breed buck, definition of, 3
Breeder
    definition of, 3
    professional, advice from, 13
Breeding, 37-50
    choosing sire, 41
    mating, 41-43

problems in, 45
stud service, 44-45
types of, 39
of young does, 45-46
Breeding season, 3, 43-44
Breeding-up
definition of, 38
methods of, 39-40
Breeds
French Alpine, 6, 48
La Mancha, 6, 50
Nubian, 6, 48-50
Saanen, 6, 15, 46
Toggenburg, 6, 46-48
Broccoli soup, recipe for, 152
Browsing, description of, 1
Brucellosis, 6, 35, 106
Bruises, care of, 117
Buck(s)
breed, 3
choosing, 25, 41
excess, 58
housing for, 94-95
odor of, 94
urinary stones in, 124-25
Bucklings, definition of, 3
Butchering, 157-58
Butter, making of, 136-38
C Calcium deficiency, in pregnant does, 120
Calcium stones, in urinary tract, 124
Calculi, in urinary tract, 124-25
Capretti lattanti alla Nonna, recipe for, 158
Caraway seed, added to cheese, 146
Carbohydrate deficiency, in pregnant does, 119
Caseous lymphadenitis. See Abscess
Cauliflower pie, recipe for, 133-34
Cauliflower soup, recipe for, 153
Cedars, as feed, 75
Celery soup, recipe for, 154
Cheese, making of, 138-44
blue, 146
as business, 177-78
chèvre, 146-48
cream, 146

herbs added to, 146
Cherry, wild, toxicity of, 117
Chevon. *See* Meat
Chèvre cheese, 146-48
China clay, for diarrhea, 118
Chives, added to cheese, 146
Chlorine, for washing teats, 31
Chokeberries, toxicity of, 75
Churn, need for, 136
Classification, of goats, 5-6
Cleanliness. *See* Sanitation
Clostridial diseases, 65
inoculation for, 106, 116, 117
Clots, in milk, 31
Clover
as forage, 74, 75
as hay, 77
Coat. *See also* Hair loss
evaluation of, 20
problems with, 68, 126
Coccidiosis, 111-12
in young kids, 62
Coliform bacteria, in milk, 35
Colostrum, 60
description of, 3
feeding to kids, 55, 57
Comfrey
as fodder, 75
tea, as fertilizer, 165
to treat bruises, 117
Commercial dairying, 173-85
Composting, of manure, 164
Compress, herbal, for udder problems, 123
Conditioning, to electric fences, 99
Containers, for grain storage, 89
Cooking, with goat's milk, 133-36, 150-56
butter, 136-38
cheese, 138-48
soups, 152-56
yogurt, 130-33
Coordination, lack of, enterotoxemia and, 118
Corn, as feed, 80
Cornbread, recipe for, 132
Corynebacterium. See Abscess
Costs, of goatkeeping, 13, 15-18.
See also Economics
Coughing, lungworms and, 110

Cranberry concentrate, for urinary stones, 125
Cream cheese, from yogurt, 146
Cream separator, need for, 136
Cud-chewing, 2, 27
Culturing agents, for yogurt, 130, 131
Curried chevon, recipe for, 159
Custard, recipe for, 150-51
Cuts, care of, 116
D Dairy Herd Improvement Association (DHIA), records of, 4-5
Dam, definition of, 4
Dehorning. *See* Disbudding
Delivery. *See* Kidding
Derris powder. *See* Rotenone
Diapers, for draining cheese, 143
Diarrhea
  coccidiosis and, 112
  enterotoxemia and, 118
  roundworms and, 110
  in young kids, 62
Diatomaceous earth, to control parasites, 67
Diet. *See* Feeding
Disbudding, 6
  of young kid, 63-66
Diseases. *See* Illness
Distention, abdominal
  overeating and, 118
  worms and, 110
Doelings, definition of, 3
Donkey, as herd company, 14
Drying off, before kidding, 52-53
E Eating habits, 1-2, 74
Echinacea coverings
  for abscess, 115
  for mastitis, 115, 124
Economics, of goatkeeping, 170-88
Edema, 7, 123
Eimeria. See Coccidiosis
Elastrators, for disbudding, 65
Elder flower
  added to milk, 156
  for herbal compress, 123
Electric fencing, 99-103
Enterotoxemia, 118-19
Epsom salts, for foot rot, 126
Estrus. *See* Heat

Evergreens, as feed, 75
Eversion, of uterus, 122
Exercise, of pregnant does, 59
F Feeding, 73-82
  fodder, 75-76
  grain, 79-82
  hay, 76-79
  of kids, 11, 55-58, 62-63
  pasture, 74-75
  pattern of, 1
  of pregnant does, 51-52
Feed storage, 89
Fences
  barbed wire, 99, 116
  electric, 99-103
  jumping of, 96-97, 98
Fennel, added to milk, 156
Fertilizers, chemical, 163
Fighting, as cause of abortion, 116
Finances. *See* Costs; Economics
Firs, as feed, 75
Fodder, plants for, 75-76
Foot rot, 126
Freezing
  of colostrum, 57
  of milk, 130
French Alpine breed, 6, 48
Freshening, 29
  definition of, 3
Fungus infections, 126
G Garden, use of manure in, 161-69
Garlic
  added to cheese, 146
  taste of, in milk, 67, 82
  for worms, 67
Gas, stomach, from overeating, 117-18
Gates, in electric fences, 100
Gestation period, 3, 28. *See also* Pregnancy
Gjetost cheese, from whey, 133
Glycerin, for pregnant does, 119
Goat(s)
  biology of, 26-29
  breed types, 6, 46-50
  cost of, 15-18
  eating habits of, 1-2, 74
  judging physical appearance of, 17, 20, 107
  manure from, 161-69

Goatkeeping
  economics of, 13, 15–18, 170–88
  failure at, reasons for, 186–88
  management, 96–104
  record-keeping in, 4–5, 19, 44,
    171
  terms of, 1–7
Goldenseal powder
  for herbal compress, 123
  for mange, 127
  to promote wound healing, 114,
    117, 126
Grade, of goats, 4
Grain, 79–82
  molding of, 117
  pelletizing of, 79, 182
  for pregnant does, 51, 59, 81
  as reward during milking, 31
  storage of, 89
Gratin dauphinois, recipe for, 135
Grazing, 84–85
  in garden, 162
Greens soup, recipe for, 153
H Haemonchus. See Worms, internal
Hair loss
  lice and, 126
  mites and, 126
  postpartum, 68, 72
Half grade, definition of, 4
Hand-rearing, of kid(s), 58
Hay, 76–79
  curing of, 78
  moldy, 117
  plants for, 77
  during winter months, 87–89
Hay loft, as insulation, 86
Health problems. See Illness
Heat
  cycle oddities, 122
  definition of, 3
  symptoms of, 38, 41–42
Hematoma, 114
Herbs
  added to cheese, 146
  for compress, 123
  to flavor milk, 156
  for mastitis, 124
Herd
  breeding, 185–86
  commercial, 173–85

  definition of, 1
  family, 171–73
  improving milk yield of, 25
  social order of, 2–3, 9, 14
Herd company, need for, 9, 14
Herd queen, definition of, 2
Hooves
  foot rot of, 126
  trimming of, 20, 68
Hormonal injections, for heat cycle
  variations, 122
Horns, 6. See also Disbudding
  problems with, 10, 65–66, 93–94
Horse(s)
  as herd company, 14
  tetanus and, 65, 116
Hydrangea root, for urinary
  stones, 125
Hydrogen peroxide, to disinfect
  cuts, 116
I Illness, 6–7, 108
  abortion, 116
  abscess, 113–15
  external parasites, 125–28
  internal parasites, 67–68,
    108–13
  mastitis, 123–24
  milk fever, 120
  from overeating, 117–19
  poisoning, 117
  from undereating, 119
  urinary stones in males, 124–25
  uterine inflammations, 122
  wounds, 116–17
Immunization. See Inoculation
Inbreeding, definition of, 39
Indigestion, from overeating, 118
Infertility, 45
In-kid. See Pregnancy
Inoculation
  against clostridial diseases, 106,
    116, 118
  against Corynebacteria, 113
  against tetanus, 65, 99, 116
Insulators, on electric fencing,
  101–3
Iodine
  for disinfecting cuts, 116
  for washing teats, 31

J Jerusalem artichokes, as fodder, 75
K Kaolin, for diarrhea, 118
Kelp, for pregnant does, 122
Ketosis, in pregnant does, 119
Kicking, during milking, 11–12
Kid(s)
  definition of, 3
  disbudding of, 63–66
  feeding of, 11, 55–58, 62–63
  postnatal care of, 61–63
  tattooing of, 66
  weaning of, 28, 63
Kidding, 29, 59–61
  care after, 61–63
  premature, 116
  presentation
    normal, 61
    unusual, 120–22
  separation after, 55–56
  signs of, 59–60
  stall for, 53–54
Kraftborner method, of drying off, 52
L Labor, human, needed in goatkeeping, 14–15, 171
Lactation, 28–29
  drying off, 52–53
  length of, 3, 28
La Mancha breed, 6, 50
Lancing, of abscess, 114
Lassie, made with yogurt, 132
Lavender, for lice, 72
Lean-to, as shelter, 84
Legume hays, 77
  overeating of, 117
Lemonbalm, added to milk, 156
Letting down
  definition of, 3
  goat's control over, 11
Lice, 68–72, 126
Lindane, for mites, 126
Linebreeding, definition of, 39
Listlessness
  ketosis and, 119
  worms and, 110
Litter, deep, during winter, 86–87
Liver flukes, 110
Locks, goats and, 96–97
Lungworms, 110
Lymph system, disease of, 113–15

M Mange, caused by mites, 126
Manure
  tea, 165
  use of, 161–69
Marinade, for goat meat, 158
Mastitis, 6, 123–24
  cider vinegar and, 52
  echinacea and, 115
Mating, 41–43
Meat, 156–58
  recipes for, 158–59
  selling bucks for, 58–59
Metritis, 122
Microorganisms, in bedding, 54
Milk
  blood in, 123
  checking for problems, 20, 31, 72
  cooking with, 133–36, 150–56
    butter, 136–38
    cheese, 138–48
    yogurt, 130–33
  flavoring of, with herbs, 156
  pasteurizing of, 36
  production during winter, 87
  storage of, 130
  taste of foodstuffs in, 67, 82
  withholding during disease treatment, 72, 111, 124, 126
Milk fever, 120
Milking, 30–36
  to extend lactation, 28–29
  before kidding, 60
  location for, 34–35, 89
  problems during, 11–12
  routine of, 34
  technique of, 30–33
    practicing of, 33–34
Milking machines, use of, 185
Milking period. *See* Lactation
Milking stand, 31, 34–35
  feeding of pregnant does on, 60
Milk replacers, feeding to kid(s), 57
Milkweed, toxicity of, 75, 117
Mineral supplements, 75–76
Mint, added to milk, 156
Miscarriage, 116
Mites, 126–27

Molasses, for pregnant does, 52, 119
Mold, on feed
    indigestion from, 118
    poisoning from, 117
Mountain laurel, toxicity of, 75, 117
Mugwort, toxicity of, 117
Mulching, along electric fence, 100
Mullein, toxicity of, 117
Musculature, of udder, 20
Myrtle, toxicity of, 75
N Nan bread, recipe for, 134
Nausea, as sign of poisoning, 117
Nematodirus roundworms, 110
Nubian breed, 6, 48–50
Nursing, effect of, on udder, 56
O Odor, of buck, 94
Omasum, 27
Omelettes, with goat cheese, 150
Onion
    added to cheese, 146
    taste of, in milk, 82
On test, definition of, 5
Oocysts, in coccidiosis, 111
Ostertagia. See Worms, internal
Outcrossing, definition of, 39
Ovaries, cystic, 122
Overeating, problems from, 117–19
P Pan feeding, of kids, 11, 58
Papers, of registration, 4
Paralysis, due to milk fever, 120
Parasites
    external, 125–28
    internal, 67–68, 108–13
    on tethered goats, 103
Parlor, milking, benefits of, 35, 89
Pasture, 90–93
    favorite plants in, 74–75, 92
    fencing of, 99–103
    rotating, 67, 109
Pattern. *See* Routine
Pedicures, 68
Pedigree, definition of, 3
Pelletizing, of feed, 79, 182
Penicillin, use on wounded animal, 116
Peppermint, added to milk, 156
Periwinkle, toxicity of, 75
Pet, goat as, 14

Physical appearance, evaluation of, 17, 20, 107
Pinbones, appearance of, before kidding, 60
Pine needles, as feed, 74
Pituitary gland, heat and, 41, 43
Pizza, recipe for, 135–36
Placenta, 61
    in compost pile, 61
    expelling after abortion, 116
    retained, 122
Plastic, for milk storage, 138
Pneumonia, 87, 107
Poisoning
    by toxic plants, 116, 117
    uremic, 125
Polled goat, definition of, 6
Pony, as herd company, 14
Postnatal care, 61–63
Potato soup, recipe for, 155
Pregnancy
    appetite during, 81
    diet for, 51–52, 81
    drying off during, 52–53
    ketosis during, 119
    length of, 3, 28
    milk fever during, 120
    signs of, 43
    stress from, 28
    worming during, 67–68
Presentation, of kid
    normal, 61
    unusual, 120–22
Production records. *See* Record-keeping
Prolapse, uterine, 122
Protein deficiency, in pregnant does, 119
Proven-out buck, definition of, 3, 41
Pulse rate, 108
Pumpkin soup, recipe for, 155
Purebreds, definition of, 4
Pyometra, 122
Pyrethrin, for lice, 72
Q Quassia chips, for lice, 72, 126
Quatre-quarts, recipe for, 151–52
Queen, of herd, 2
Quiche Lorraine, recipe for, 148–49
R Raccoons, in grain, 89

Raised-bed gardening, 168–69
Raspberry leaves, for pregnant
    does, 122
Ratatouille, 156
Recorded grade, definition of, 4
Record-keeping
    in breeding, 44
    in family herd, 171
    production, 5, 19
    in registration, 4–5
Registered stock, 3–4
    versus scrubs, 13–14
Registration, of goats, 3–4
Rennet, for cheesemaking, 140
    French, 147
Reproduction. *See* Heat;
    Pregnancy
Respiration rate, 108
Respiratory difficulties, worms
    and, 110
Resting, between pregnancies, 28
Reticulum, 27
Reward, during milking, 31
Rhododendron, toxicity of, 75, 117
Rice pudding, recipe for, 151
Ringworm, 126
Rodents, in grain, 89
Root crop soup, recipe for, 155
Rotenone
    for lice, 72, 126
    for ticks, 127
Roundworms, 108–9
Routine, goats' need for, 34, 98
Rubber bands, for disbudding, 65
Rue, for worms, 67
Rumen, action of, 2, 27
Ruminants, description of, 27–28
Rutting season. *See* Breeding
    season
Ryegrass, as feed, 74, 75
S Saanen breed, 6, 15, 46
Sage, added to cheese, 146
Saint-John's-wort, toxicity of, 117
Salad dressing, yogurt as, 132
Salivation, abnormal, as sign of
    poisoning, 117
Salt, in butter, 137–38
Salt block, 76
Sanitation
    of kidding stall, 54

of milking area, 35
    of stalls, 164
Scotch hands, to work butter, 138
Scours, in young kids, 62
Scratches, care of, 117
Scrub
    description of, 3
    evaluation of, 19–20
    versus registered stock, 13–14
Scur, growth of, after disbudding,
    63
Season. *See* Breeding season
Seaweed emulsions, as fertilizer,
    165
Separation, of dam and kid(s),
    55–56
Septicemia, after abortion, 116
Shade, need for, 84
Sheds, as shelter, 84–87
Shelter, 15, 83–89
    for buck, 94–95
Sire
    choosing of, 41
    definition of, 4
Skimming spoon, for butter-
    making, 137
Skin, postpartum problems with,
    72
Slaughtering, of meat animal, 157
Slope, in stall, to reposition fetus,
    120
Social order, of herd, 2–3, 9, 14
Soufflé, recipe for, 149–50
Soups, recipes for, 152–56
Splicing, of electric fencing, 101
Spring rise, of parasite
    populations, 108
Spruces, as feed, 75
Stalls, separate, 93–94
    for kidding, 53–54
Stand, for milking. *See* Milking
    stand
Star certificates, 4, 5
Stomach, of goat, 2, 27
Stones, in urinary tract, 124–25
Stool. *See also* Diarrhea
    bloody, 112
    worms and, 110
    of young kids, 62

INDEX

Stress, as cause of abortion, 116
Strip cup, for checking milk, 31, 72
Stud service, arranging for, 44-45
Stud service memo, 44
Sulfa ointment, for mites, 127
Sulfonamides, for coccidiosis, 111
Sunflowers, as fodder, 75
Sunlight
    antibacterial action of, 62, 114, 117
    as fungicidal agent, 126
    lice and, 72
    pregnant does and, 59
    shade from, 84
Supplements
    mineral, 75-76
    vitamin, 72

T Tamari marinade, 158
Tattooing, of kids, 66
Teats, washing of, 31, 124
Temperature, body. *See* Body temperature
Terms, of goatkeeping, 1-7
Testing, regular
    for disease, 35, 106
    for internal parasites, 67, 109
Tetanus, inoculation against, 65, 99, 116
Tethering, 92, 103-4
Tetramisole, as wormer, 110, 111
Thiabendazole, for worming, 68, 111
Three-quarter grade, definition of, 4
Thyme, added to cheese, 146
Ticks, 127
Tilsit cheese, 146
Toggenburg breed, 6, 46-48
Tomato soup, recipe for, 154
Toxic plants, 75, 117
    as cause of abortion, 116
Transporting, of goats, 10
Trichostrongulus. See Worms, internal
Tuberculosis, 6, 35, 106

U Udder
    appearance of
        during drying off, 52-53
        before kidding, 60
    effect of nursing on, 56
    evaluation of, 20
    problems with, 6-7, 53, 123
        mastitis, 6, 52, 115, 123-24
Umbilical cord, care of, 62
Undereating, problems from, 119
Urea, in grain, 80
Uremic poisoning, 125
Urinary calculi, in bucks, 124-25
Uterus
    inflammation of, 122
    prolapse of, 122

V Vaccines. *See* Inoculation
Vegetable oil, for indigestion, 118
Vegetable rennet, for cheesemaking, 140
Veterinarian, finding of, 106
Vinca minor, toxicity of, 75
Vinegar, apple cider
    for pregnant does, 52
    for urinary stones, 125
Vitamin B_{12} deficiency, 72
Vitamin deficiencies, postpartum, 72
Vitamin E, to promote healing, 114, 117, 126
Vomiting, as sign of poisoning, 117

W Washing. *See also* Sanitation
    of teats, 31, 124
Weaning, of kid, 28, 63
Weight loss, worms and, 109
Wethers
    definition of, 3
    raising for meat, 156-57
Wheat germ oil, for newly delivered does, 72
Whey, use of, 133
Whitewashing, of kidding stall, 54
Winter, shelter during, 86-87
Wire, barbed. *See* Barbed wire
Worms, internal, 67-68, 81, 108-13
Wormwood
    for urinary stones, 125
    for worms, 67
Wounds, care of, 116

Y Yew, toxicity of, 117
Yogurt, 130-33
    cream cheese from, 146

Z Zucchini soup, recipe for, 154